CURIOSITIES OF SCIENCE.

The plan of this work is thus sketched in the *Introduction*:

"There have been in the history of Art, four grand styles of imitating Nature—Tempera, Encaustic, Fresco, and Oil. These, together with the minor modes of Painting, we propose arranging in something like chronological sequence; but our design being to offer an explanation of the Art derived from practical acquaintance, rather than attempt to give its history, we shall confine ourselves for the most part to so much only of the History of Painting as is necessary to elucidate the origin of the different practices which have obtained at different periods."

By this means, the Authors hope to produce a work which may be valuable to the Amateur, and interesting to the Connoisseur, the Artist, and the General Reader.

TO THE READER

GENTLE READER ,

The volume of " CURIOSITIES " which I here present to your notice is a portion of the result of a long course of reading, observation, and research, necessary for the compilation of thirty volumes of "Arcana of Science" and "Year-Book of Facts," published from 1828 to 1858. Throughout this period—nearly half of the Psalmist's "days of our years"—I have been blessed with health and strength to produce these volumes, year by year (with one exception), upon the appointed day; and this with unbroken attention to periodical duties, frequently rendered harassing or ungenial. Nevertheless, during these three decades I have found my account in the increasing approbation of the reading public, which has been so largely extended to the series of " THINGS NOT GENERALLY KNOWN ," of which the present volume of " CURIOSITIES OF SCIENCE " is an instalment. I need scarcely add, that in its progressive preparation I have endeavoured to compare, weigh, and consider, the contents, so as to combine the experience of the Past with the advantages of the Present.

In these days of universal attainments, when Science becomes not merely a luxury to the rich, but bread to the poor, and when the very amusements as well as the conveniences of life have taken a scientific colour, it is reasonable to hope that the present volume may be acceptable to a large class of seekers after "things not generally known." For this purpose, I have aimed at soundness as well as popularity; although, for myself, I can claim little beyond being one of those industrious "ants of science" who garner facts, and by selection and comparison adapt them for a wider circle of readers than they were originally expected to reach. In each case, as far as possible, these " CURIOSITIES " bear the mint-mark of authority; and in the living list are prominent the names of Humboldt and Herschel, Airy and Whewell, Faraday, Brewster, Owen, and Agassiz, Maury, Wheatstone, and Hunt, from whose writings and researches the following pages are frequently enriched.

The sciences here illustrated are, in the main, Astronomy and Meteorology; Geology and Paleontology; Physical Geography; Sound, Light, and Heat; Magnetism and Electricity,—the latter with special attention to the great marvel of our times, the Electro-magnetic Telegraph. I hope, at no very distant period, to extend the "CURIOSITIES" to another volume, to include branches of Natural and Experimental Science which are not here presented.

I. T.

November 1858.

The Frontispiece.

THE GREAT ROSSE TELESCOPE.

The originator and architect of this magnificent instrument had long been distinguished in scientific research as Lord Oxmantown; and may be considered to have gracefully commemorated his succession to the Earldom of Rosse, and his Presidency of the Royal Society, by the completion of this marvellous work, with which his name will be hereafter indissolubly associated.

The Great Reflecting Telescope at Birr Castle (of which the Frontispiece represents a portion1) will be found fully described at pp. 96–99 of the present volume of *Curiosities of Science.*

This matchless instrument has already disclosed "forms of stellar arrangement indicating modes of dynamic action never before contemplated in celestial mechanics." "In these departments of research,—the examination of the configurations of nebulæ, and the resolution of nebulæ into stars (says the Rev. Dr. Scoresby),—the six-feet speculum has had its grandest triumphs, and the noble artificer and observer the highest rewards of his talents and enterprise. Altogether, the quantity of work done during a period of about seven years—including a winter when a noble philanthropy for a starving population absorbed the keenest interests of science—has been decidedly great; and the new knowledge acquired concerning the handiwork of the great Creator amply satisfying of even sanguine expectation."

The Vignette.

SIR HUMPHRY DAVY'S OWN MODEL OF HIS SAFETY-LAMP.

Of the several contrivances which have been proposed for safely lighting coal-mines subject to the visitation of fire-damp, or carburetted hydrogen, the Safety-Lamp of Sir Humphry Davy is the only one which has ever been judged safe, and been extensively employed. The inventor first turned his attention to the subject in 1815, when Davy began a minute chemical examination of fire-damp, and found that it required an admixture of a large quantity of atmospheric air to render it explosive. He then ascertained that explosions of inflammable gases were incapable of being passed through long narrow metallic tubes, and that this principle of security was still obtained by diminishing their length and increasing their number. This fact led to trials upon sieves made of wire-gauze; when Davy found that if a piece of wire-gauze was held over the flame of a lamp, or of coal-gas, it prevented the flame from passing; and he ascertained that a flame confined in a cylinder of very fine wire-gauze did not explode even in a mixture of oxygen and hydrogen, but that the gases burnt in it with great vivacity.

These experiments served as the basis of the Safety-Lamp. The apertures in the gauze, Davy tells us in his work on the subject, should not be more than 1/22d of an inch square. The lamp is screwed on to the bottom of the wire-gauze cylinder. When it is lighted, and gradually introduced into an atmosphere mixed with fire-damp, the size and length of the flame are first increased. When the inflammable gas forms as much as 1/12th of the volume of air, the cylinder becomes filled with a feeble blue flame, within which the flame of the wick burns brightly, and the light of the wick continues till the fire-damp increases to 1/6th or 1/5th; it is then lost in the flame of the fire-damp, which now fills the cylinder with a pretty strong light; and when the foul air constitutes one-third of the atmosphere it is no longer fit for respiration,—and this ought to be a signal to the miner to leave that part of the workings.

Sir Humphry Davy presented his first communication respecting his discovery of the Safety-Lamp to the Royal Society in 1815. This was followed by a series of papers remarkable for their simplicity and clearness, crowned by that read on the 11th of January 1816, when the principle of the Safety-Lamp was announced, and Sir Humphry presented to the Society a model made by his own hands, which is to this day preserved in the collection of the Royal Society at Burlington House. From this interesting memorial the Vignette has been sketched.

There have been several modifications of the Safety-Lamp, and the merit of the discovery has been claimed by others, among whom was Mr. George Stephenson; but the question was set at rest forty-one years since by an examination,—attested by Sir Joseph Banks, P.R.S., Mr. Brande, Mr. Hatchett, and Dr. Wollaston,—and awarding the independent merit to Davy.

A more substantial, though not a more honourable, testimony of approval was given by the coal-owners, who subscribed 2500*l*. to purchase a superb service of plate, which was suitably inscribed and presented to Davy.2

Meanwhile the Report by the Parliamentary Committee "cannot admit that the experiments (made with the Lamp) have any tendency to detract from the character of Sir Humphry Davy, or to disparage the fair value placed by himself upon his invention. The improvements are probably those which longer life and additional facts would have induced him to contemplate as desirable, and of which, had he not been the inventor, he might have become the patron."

The principle of the invention may be thus summed up. In the Safety-Lamp, the mixture of the fire-damp and atmospheric air within the cage of wire-gauze explodes upon coming in contact with the flame; but the combustion cannot pass through the wire-gauze, and being there imprisoned, cannot impart to the explosive atmosphere of the mine any of its force. This effect has been erroneously attributed to a cooling influence of the metal.

Professor Playfair has eloquently described the Safety-Lamp of Davy as a present from philosophy to the arts; a discovery in no degree the effect of accident or chance, but the result of patient and enlightened research, and strongly exemplifying the great use of an immediate and constant appeal to experiment. After characterising the invention as the *shutting-up in a net of the most slender texture* a most violent and irresistible force, and a power that in its tremendous effects seems to emulate the lightning and the earthquake, Professor Playfair thus concludes: "When to this we add the beneficial consequences, and the saving of the lives of men, and consider that the effects are to remain as long as coal continues to be dug from the bowels of the earth, it may be fairly said that there is hardly in the whole compass of art or science a single invention of which one would rather wish to be the author.... This," says Professor Playfair, "is exactly such a case as we should choose to place before Bacon, were he to revisit the earth; in order to give him, in a small compass, an idea of the advancement which philosophy has made since the time when he had pointed out to her the route which she ought to pursue."

Introductory.

SCIENCE OF THE ANCIENT WORLD.

In every province of human knowledge where we now possess a careful and coherent interpretation of nature, men began by attempting in bold flights to leap from obvious facts to the highest point of generality—to some wide and simple principle which after-ages had to reject. Thus, from the facts that all bodies are hot or cold, moist or dry, they leapt at once to the doctrine that the world is constituted of four elements—earth, air, fire, water; from the fact that the heavenly bodies circle the sky in courses which occur again and again, they at once asserted that they move in exact circles, with an exactly uniform motion; from the fact that heavy bodies fall through the air somewhat faster than light ones, it was assumed that all bodies fall quickly or slowly exactly in proportion to their weight; from the fact that the magnet attracts iron, and that this force of attraction is capable of increase, it was inferred that a perfect magnet would have an irresistible force of attraction, and that the magnetic pole of the earth would draw the nails out of a ship's bottom which came near it; from the fact that some of the finest quartz crystals are found among the snows of the Alps, it was inferred that the crystallisation of gems is the result of intense and long-continued cold: and so on in innumerable instances. Such anticipations as these constituted the basis of almost all the science of the ancient world; for such principles being so assumed, consequences were drawn from them with great ingenuity, and systems of such deductions stood in the place of science.—*Edinburgh Review*, No. 216.

SCIENCE AT OXFORD AND CAMBRIDGE.

The earliest science of a decidedly English school is due, for the most part, to the University of Oxford, and specially to Merton College,—a foundation of which Wood remarks, that there was no other for two centuries, either in Oxford or Paris, which could at all come near it in the cultivation of the sciences. But he goes on to say that large chests full of the writers of this college were allowed to remain untouched by their successors for fear of the magic which was supposed to be contained in them. Nevertheless, it is not difficult to trace the liberalising effect of scientific study upon the University in general, and Merton College in particular; and it must be remembered that to the cultivation of the mind at Oxford we owe almost all the literary celebrity of the middle ages. In this period the University of Cambridge appears to have acquired no scientific distinction. Taking as a test the acquisition of celebrity on the continent, we find that Bacon, Sacrobosco, Greathead, Estwood, &c. were all of Oxford. The latter University had its morning of splendour while Cambridge was comparatively unknown; it had also its noonday, illustrated by such men as Briggs, Wren, Wallis, Halley, and Bradley.

The age of science at Cambridge may be said to have begun with Francis Bacon; and but that we think much of the difference between him and his celebrated namesake lies more in time and circumstances than in talents or feelings, we would rather date from 1600 with the former than from 1250 with the latter. Praise or blame on either side is out of the question, seeing that the earlier foundation of Oxford, and its superiority in pecuniary means, rendered all that took place highly probable; and we are in a great measure indebted for the liberty of writing our thoughts, to the cultivation of the liberalising sciences at Oxford in the dark ages.

With regard to the University of Cambridge, for a long time there hardly existed the materials of any proper instruction, even to the extent of pointing out what books should be read by a student desirous of cultivating astronomy.

PLATO'S SURVEY OF THE SCIENCES.

Plato, like Francis Bacon, took a review of the sciences of his time: he enumerates arithmetic and plane geometry, treated as collections of abstract and permanent truths; solid geometry, which he "notes as deficient" in his time, although in fact he and his school were in possession of the doctrine of the "five regular solids;" astronomy, in which he demands a science which should be elevated above the mere knowledge of phenomena. The visible appearances of the heavens only suggest the problems with which true astronomy deals; as beautiful geometrical diagrams do not prove, but only suggest geometrical propositions. Finally, Plato notices the subject of harmonics, in which he requires a science which shall deal with truths more exact than the ear can establish, as in astronomy he requires truths more exact than the eye can assure us of.

In a subsequent paper Plato speaks of *Dialectic* as a still higher element of a philosophical education, fitted to lead men to the knowledge of real existences and of the supreme good. Here he describes dialectic by its objects and purpose. In other places dialectic is spoken of as a method or process of analysis; as in the *Phædrus*, where Socrates describes a good dialectician as one who can divide a subject according to its natural members, and not miss the joint, like a bad carver. Xenophon says that Socrates derived *dialectic* from a term implying to *divide a subject into parts*, which Mr. Grote thinks unsatisfactory as an etymology, but which has indicated a practical connection in the Socratic school. The result seems to be that Plato did not establish any method of analysis of a subject as his dialectic; but he conceived that the analytical habits formed by the comprehensive study of the exact sciences, and sharpened by the practice of dialogue, would lead his students to the knowledge of first principles.—*Dr. Whewell.*

FOLLY OF ATHEISM.

Morphology, in natural science, teaches us that the whole animal and vegetable creation is formed upon certain fundamental types and patterns, which can be traced under various modifications and transformations through all the rich variety of things apparently of most dissimilar build. But here and there a scientific person takes it into his foolish head that there may be a set of moulds without a moulder, a calculated gradation of forms without a calculator, an ordered world without an ordering God. Now, this atheistical science conveys about as much meaning as suicidal life: for science is possible only where there are ideas, and ideas are only possible where there is mind, and minds are the offspring of God; and atheism itself is not merely ignorance and stupidity,—it is the purely nonsensical and the unintelligible.— *Professor Blackie*; *Edinburgh Essays*, 1856.

THE ART OF OBSERVATION.

To observe properly in the very simplest of the physical sciences requires a long and severe training. No one knows this so feelingly as the great discoverer. Faraday once said, that he always doubts his own observations. Mitscherlich on one occasion remarked to a man of science that it takes fourteen years to discover and establish a single new fact in chemistry. An enthusiastic student one day betook himself to Baron Cuvier with the exhibition of a new organ—a muscle which he supposed himself to have discovered in the body of some living creature or other; but the experienced and sagacious naturalist kindly bade the young man return to him with the same discovery in six months. The Baron would not even listen to the student's demonstration, nor examine his dissection, till the eager and youthful discoverer had hung over the object of inquiry for half a year; and yet that object was a mere thing of the senses.—*North-British Review*, No. 18.

MUTUAL RELATIONS OF PHENOMENA.

In the observation of a phenomenon which at first sight appears to be wholly isolated, how often may be concealed the germ of a great discovery! Thus, when Galvani first stimulated the nervous fibre of the frog by the accidental contact of two heterogeneous metals, his contemporaries could never have anticipated that the action of the voltaic pile would discover to us in the alkalies metals of a silver lustre, so light as to swim on water, and eminently inflammable; or that it would become a powerful instrument of chemical analysis, and at the same time a thermoscope and a magnet. When Huyghens first observed, in 1678, the phenomenon of the polarisation of light, exhibited in the difference between two rays into which a pencil of light divides itself in passing through a doubly refracting crystal, it could not have been foreseen that a century and a half later the great philosopher Arago would, by his discovery of *chromatic polarisation*, be led to discern, by means of a small fragment of Iceland spar, whether solar light emanates from a solid body or a gaseous covering; or whether comets transmit light directly, or merely by reflection.—*Humboldt's Cosmos*, vol. i.

PRACTICAL RESULTS OF THEORETICAL SCIENCE.

What are the great wonders, the great sources of man's material strength, wealth, and comfort in modern times? The Railway, with its mile-long trains of men and merchandise, moving with the velocity of the wind, and darting over chasms a thousand feet wide; the Electric Telegraph, along which man's thoughts travel with the velocity of light, and girdle the earth more quickly than Puck's promise to his master; the contrivance by which the Magnet, in the very middle of a strip of iron, is still true to the distant pole, and remains a faithful guide to the mariner; the Electrotype process, by which a metallic model of any given object, unerringly exact, grows into being like a flower. Now, all these wonders are the result of recent and profound discoveries in theoretical science. The Locomotive Steam-engine, and the Steam-engine in all its other wonderful and invaluable applications, derives its efficacy from the discoveries, by Watt and others, of the laws of steam. The Railway Bridge is not made strong by mere accumulation of materials, but by the most exact and careful scientific examination of the means of giving the requisite strength to every part, as in the great example of Mr. Stephenson's Britannia Bridge over the Menai Strait. The Correction of the Magnetic Needle in iron ships it would have been impossible for Mr. Airy to secure without a complete theoretical knowledge of the laws of Magnetism. The Electric Telegraph and the Electrotype process include in their principles and mechanism the most complete and subtle results of electrical and magnetical theory.—*Edinburgh Review*, No. 216.

PERPETUITY OF IMPROVEMENT.

In the progress of society all great and real improvements are perpetuated: the same corn which, four thousand years ago, was raised from an improved grass by an inventor worshiped for two thousand years in the ancient world under the name of Ceres, still forms the principal food of mankind; and the potato, perhaps the greatest benefit that the old has derived from the new world, is spreading over Europe, and will continue to nourish an extensive population when the name of the race by whom it was first cultivated in South America is forgotten.—*Sir H. Davy.*

THE EARLIEST ENGLISH SCIENTIFIC TREATISE.

Geoffrey Chaucer, the poet, wrote a treatise on the Astrolabe for his son, which is the earliest English treatise we have met with on any scientific subject. It was not completed; and the apologies which Chaucer makes to his own child for writing in English are curious; while his inference that his son should therefore "pray God save the king that is lord of this language," is at least as loyal as logical.

PHILOSOPHERS' FALSE ESTIMATES OF THEIR OWN LABOURS.

Galileo was confident that the most important part of his contributions to the knowledge of the solar system was his Theory of the Tides—a theory which all succeeding astronomers have rejected as utterly baseless and untenable. Descartes probably placed far above his beautiful explanation of the rainbow, his *à priori* theory of the existence of the vortices which caused the motion of the planets and satellites. Newton perhaps considered as one of the best parts of his optical researches his explanation of the natural colour of bodies, which succeeding optical philosophers have had to reject; and he certainly held very strongly the necessity of a material cause for gravity, which his disciples have disregarded. Davy looked for his greatest triumph in the application of his discoveries to prevent the copper bottoms of ships from being corroded. And so in other matters.—*Edinburgh Review*, No. 216.

RELICS OF GENIUS.

Professor George Wilson, in a lecture to the Scottish Society of Arts, says: "The spectacle of these things ministers only to the good impulses of humanity. Isaac Newton's telescope at the Royal Society of London; Otto Guericke's air-pump in the Library at Berlin; James Watt's repaired Newcomen steam-engine in the Natural-Philosophy class-room of the College at Glasgow; Fahrenheit's thermometer in the corresponding class-room of the University of Edinburgh; Sir H. Davy's great voltaic battery at the Royal Institution, London, and his safety-lamp at the Royal Society; Joseph Black's pneumatic trough in Dr. Gregory's possession; the first wire which Faraday made rotate electro-magnetically, at St. Bartholomew's Hospital; Dalton's atomic models at Manchester; and Kemp's liquefied gases in the Industrial Museum of Scotland,—are alike personal relics, historical monuments, and objects of instruction, which grow more and more precious every year, and of which we never can have too many."

THE ROYAL SOCIETY: THE NATURAL AND SUPERNATURAL.

The Royal Society was formed with the avowed object of increasing knowledge by direct experiment; and it is worthy of remark, that the charter granted by Charles II. to this celebrated institution declares that its object is the extension of natural knowledge, as opposed to that which is supernatural.

Dr. Paris (*Life of Sir H. Davy*, vol. ii. p. 178) says: "The charter of the Royal Society states that it was established for the improvement of *natural* science. This epithet *natural* was originally intended to imply a meaning, of which very few persons, I believe, are aware. At the period of the establishment of the society, the arts of witchcraft and divination were very extensively encouraged; and the word *natural* was therefore introduced in contradistinction to *supernatural*."

THE PHILOSOPHER BOYLE.

After the death of Bacon, one of the most distinguished Englishmen was certainly Robert Boyle, who, if compared with his contemporaries, may be said to rank immediately below Newton, though of course very inferior to him as an original thinker. Boyle was the first who instituted exact experiments into the relation between colour and heat; and by this means not only ascertained some very important facts, but laid a foundation for that union between optics and thermotics, which, though not yet completed, now merely waits for some great philosopher to strike out a generalisation large enough to cover both, and thus fuse the two sciences into a single study. It is also to Boyle, more than to any other Englishman, that we owe the science of hydrostatics in the state in which we now possess it.3 He is also the original discoverer of that beautiful law, so fertile in valuable results, according to which the elasticity of air varies as its density. And, in the opinion of one of the most eminent modern naturalists, it was Boyle who opened up those chemical inquiries which went on accumulating until, a century later, they supplied the means by which Lavoisier and his contemporaries fixed the real basis of chemistry, and enabled it for the first time to take its proper stand among those sciences that deal with the external world.— *Buckle's History of Civilization*, vol. i.

SIR ISAAC NEWTON'S ROOMS AND LABORATORY IN TRINITY COLLEGE, CAMBRIDGE.

Of the rooms occupied by Newton during his early residence at Cambridge, it is now difficult to settle the locality. The chamber allotted to him as Fellow, in 1667, was "the Spiritual Chamber," conjectured to have been the ground-room, next the chapel, but it is not certain that he resided there. The rooms in which he lived from 1682 till he left Cambridge, are in the north-east corner of the great court, on the first floor, on the right or north of the gateway or principal entrance to the college. His laboratory, as Dr. Humphrey Newton tell us, was "on the left end of the garden, near the east end of the chapel; and his telescope (refracting) was five feet long, and placed at the head of the stairs, going down into the garden."4 The east side of Newton's rooms has been altered within the last fifty years: Professor Sedgwick, who came up to college in 1804, recollects a wooden room, supported on an arcade, shown in Loggan's view, in place of which arcade is now a wooden wall and brick chimney.

Dr. Humphrey Newton relates that in college Sir Isaac very rarely went to bed till two or three o'clock in the morning, sometimes not till five or six, especially at spring and fall of the leaf, when he used to employ about six weeks in his laboratory, the fire scarcely going out either night or day; he sitting up one night, and Humphrey another, till he had finished his chemical experiments. Dr. Newton describes the laboratory as "well furnished with chymical materials, as bodyes, receivers, heads, crucibles, &c., which was made very little use of, ye crucibles excepted, in which he fused his metals: he would sometimes, though very seldom, look into an old mouldy book, which lay in his laboratory; I think it was titled *Agricola de Metallis*, the transmuting of metals being his chief design, for which purpose antimony was a great ingredient." "His brick furnaces, *pro re nata*, he made and altered himself without troubling a bricklayer." "What observations he might make with his telescope, I know not, but several of his observations about comets and the planets may be found scattered here and there in a book intitled *The Elements of Astronomy*, by Dr. David Gregory."[5]

NEWTON'S "APPLE-TREE."

Curious and manifold as are the trees associated with the great names of their planters, or those who have sojourned in their shade, the Tree which, by the falling of its fruit, suggested to Newton the idea of Gravity, is of paramount interest. It appears that, in the autumn of 1665, Newton left his college at Cambridge for his paternal home at Woolsthorpe. "When sitting alone in the garden," says Sir David Brewster, "and speculating on the power of gravity, it occurred to him, that as the same power by which the apple fell to the ground was not sensibly diminished at the greatest distance from the centre of the earth to which we can reach, neither at the summits of the loftiest spires, nor on the tops of the highest mountains, it might extend to the moon and retain her in her orbit, in the same manner as it bends into a curve a stone or a cannon-ball when projected in a straight line from the surface of the earth."—*Life of Newton*, vol. i. p. 26. Sir David Brewster notes, that neither Pemberton nor Whiston, who received from Newton himself his first ideas of gravity, records this story of the falling apple. It was mentioned, however, to Voltaire by Catherine Barton, Newton's niece; and to Mr. Green by Martin Folkes, President of the Royal Society. Sir David Brewster saw the reputed apple-tree in 1814, and brought away a portion of one of its roots. The tree was so much decayed that it was cut down in 1820, and the wood of it carefully preserved by Mr. Turnor, of Stoke Rocheford.

De Morgan (in *Notes and Queries*, 2d series, No. 139, p. 169) questions whether the fruit was an apple, and maintains that the anecdote rests upon very slight authority; more especially as the idea had for many years been floating before the minds of physical inquirers; although Newton cleared away the confusions and difficulties which prevented very able men from proceeding beyond conjecture, and by this means established *universal* gravitation.

NEWTON'S "PRINCIPIA."

"It may be justly said," observes Halley, "that so many and so valuable philosophical truths as are herein discovered and put past dispute were never yet owing to the capacity and industry of any one man." "The importance and generality of the discoveries," says Laplace, "and the immense number of original and profound views, which have been the germ of the most brilliant theories of the philosophers of this

(18th) century, and all presented with much elegance, will ensure to the work on the *Mathematical Principles of Natural Philosophy* a preëminence above all the other productions of human genius."

DESCARTES' LABOURS IN PHYSICS.

The most profound among the many eminent thinkers France has produced, is Réné Descartes, of whom the least that can be said is, that he effected a revolution more decisive than has ever been brought about by any other single mind; that he was the first who successfully applied algebra to geometry; that he pointed out the important law of the sines; that in an age in which optical instruments were extremely imperfect, he discovered the changes to which light is subjected in the eye by the crystalline lens; that he directed attention to the consequences resulting from the weight of the atmosphere; and that he moreover detected the causes of the rainbow. At the same time, and as if to combine the most varied forms of excellence, he is not only allowed to be the first geometrician of the age, but by the clearness and admirable precision of his style, he became one of the founders of French prose. And, although he was constantly engaged in those lofty inquiries into the nature of the human mind, which can never be studied without wonder, he combined with them a long course of laborious experiment upon the animal frame, which raised him to the highest rank among the anatomists of his time. The great discovery made by Harvey of the Circulation of the Blood was neglected by most of his contemporaries; but it was at once recognised by Descartes, who made it the basis of the physiological part of his work on man. He was likewise the discoverer of the lacteals by Aselli, which, like every great truth yet laid before the world, was at its first appearance, not only disbelieved, but covered with ridicule.—*Buckle's History of Civilization*, vol. i.

CONIC SECTIONS.

If a cone or sugar-loaf be cut through in certain directions, we shall obtain figures which are termed conic sections: thus, if we cut through a sugar-loaf parallel to its base or bottom, the outline or edge of the loaf where it is cut will be *a circle*. If the cut is made so as to slant, and not be parallel to the base of the loaf, the outline is an *ellipse*, provided the cut goes quite through the sides of the loaf all round; but if it goes slanting, and parallel to the line of the loaf's side, the outline is a *parabola*, a conic section or curve, which is distinguished by characteristic properties, every point of it bearing a certain fixed relation to a certain point within it, as the circle does to its centre.—*Dr. Paris's Notes to Philosophy in Sport, &c.*

POWER OF COMPUTATION.

The higher class of mathematicians, at the end of the seventeenth century, had become excellent computers, particularly in England, of which Wallis, Newton, Halley, the Gregorys, and De Moivre, are splendid examples. Before results of extreme exactness had become quite familiar, there was a gratifying sense of power in bringing out the new methods. Newton, in one of his letters to Oldenburg, says that he was at one time too much attached to such things, and that he should be ashamed to say to what number of figures he was in the habit of carrying his results. The growth of power of computation on the Continent did not, however, keep pace with that of the same in England. In 1696, De Laguy, a well-known writer on algebra, and a member of the Academy of Sciences, said that the most skilful computer could not, in less than a month, find within a unit the cube root of 696536483318640035073641037.—*De Morgan.*

"THE SCIENCE OF THE COSMOS."

Humboldt, characterises this "uncommon but definite expression" as the treating of "the assemblage of all things with which space is filled, from the remotest nebulæ to the climatic distribution of those delicate tissues of vegetable matter which spread a variegated covering over the surface of our rocks." The word *cosmos*, which primitively, in the Homeric ages, indicated an idea of order and harmony, was subsequently adopted in scientific language, where it was gradually applied to the order observed in the movements of the heavenly bodies; to the whole universe; and then finally to the world in which this harmony was reflected to us.

Physical Phenomena.

ALL THE WORLD IN MOTION.

Humboldt, in his *Cosmos*,6 gives the following beautiful illustrative proofs of this phenomenon:

If, for a moment, we imagine the acuteness of our senses preternaturally heightened to the extreme limits of telescopic vision, and bring together events separated by wide intervals of time, the apparent repose which reigns in space will suddenly vanish; countless stars will be seen moving in groups in various directions; nebulæ wandering, condensing, and dissolving like cosmical clouds; the milky way breaking up in parts, and its veil rent asunder. In every point of the celestial vault we shall recognise the dominion of progressive movement, as on the surface of the earth where vegetation is constantly putting forth its leaves and buds, and unfolding its blossoms. The celebrated Spanish botanist, Cavanilles, first conceived the possibility of "seeing grass grow," by placing the horizontal micrometer wire of a telescope, with a high magnifying power, at one time on the point of a bamboo shoot, and at another on the rapidly unfolding flowering stem of an American aloe; precisely as the astronomer places the cross of wires on a culminating star. Throughout the whole life of physical nature—in the organic as in the sidereal world—existence, preservation, production, and development, are alike associated with motion as their essential condition.

THE AXIS OF ROTATION.

It is remarkable as a mechanical fact, that nothing is so permanent in nature as the Axis of Rotation of any thing which is rapidly whirled. We have examples of this in every-day practice. The first is the motion of *a boy's hoop*. What keeps the hoop from falling?—It is its rotation, which is one of the most complicated subjects in mechanics.

Another thing pertinent to this question is, *the motion of a quoit*. Every body who ever threw a quoit knows that to make it preserve its position as it goes through the air, it is necessary to give it a whirling motion. It will be seen that while whirling, it preserves its plane, whatever the position of the plane may be, and however it may be inclined to the direction in which the quoit travels. Now, this has greater analogy with the motion of the earth than any thing else.

Another illustration is *the motion of a spinning top*. The greatest mathematician of the last century, the celebrated Euler, has written a whole book on the motion of a top, and his Latin treatise*De motu Turbinis* is one of the most remarkable books on mechanics. The motion of a top is a matter of the greatest importance; it is applicable to the elucidation of some of the greatest phenomena of nature. In all these instances there is this wonderful tendency in rotation to preserve the axis of rotation unaltered.—*Prof. Airy's Lect. on Astronomy.*

THE EARTH'S ANNUAL MOTION.

In conformity with the Copernican view of our system, we must learn to look upon the sun as the comparatively motionless centre about which the earth performs an annual elliptic orbit of the dimensions and excentricity, and with a velocity, regulated according to a certain assigned law; the sun occupying one of the foci of the ellipse, and from that station quietly disseminating on all sides its light and heat; while the earth travelling round it, and presenting itself differently to it at different times of the year and day, passes through the varieties of day and night, summer and winter, which we enjoy.—*Sir John Herschel's Outlines of Astronomy.*

Laplace has shown that the length of the day has not varied the hundredth part of a second since the observations of Hipparchus, 2000 years ago.

STABILITY OF THE OCEAN.

In submitting this question to analysis, Laplace found that the *equilibrium of the ocean is stable if its density is less than the mean density of the earth*, and that its equilibrium cannot be subverted unless these two densities are equal, or that of the earth less than that of its waters. The experiments on the attraction of Schehallien and Mont Cenis, and those made by Cavendish, Reich, and Baily, with balls of lead, demonstrate that the mean density of the earth is at least *five* times that of water, and hence the stability of the ocean is placed beyond a doubt. As the seas, therefore, have at one time covered continents which are now raised above their level, we must seek for some other cause of it than any want of stability in the equilibrium of the ocean. How beautifully does this conclusion illustrate the language of Scripture, "Hitherto shalt thou come, but no further"! (*Job* xxxviii. 11.)

COMPRESSION OF BODIES.

Sir John Leslie observes, that *air compressed* into the fiftieth part of its volume has its elasticity fifty times augmented: if it continued to contract at that rate, it would, from its own incumbent weight, acquire the density of water at the depth of thirty-four miles. But water itself would have its density doubled at the depth of ninety-three miles, and would attain the density of quicksilver at the depth of 362 miles. In descending, therefore, towards the centre, through nearly 4000 miles, the condensation of ordinary substances would surpass the utmost powers of conception. Dr. Young says, that steel would be compressed into one-fourth, and stone into one-eighth, of its bulk at the earth's centre.—*Mrs. Somerville.*

THE WORLD IN A NUTSHELL.

From the many proofs of the non-contact of the atoms, even in the most solid parts of bodies; from the very great space obviously occupied by pores—the mass having often no more solidity than a heap of empty boxes, of which the apparently solid parts may still be as porous in a second degree and so on; and from the great readiness with which light passes in all directions through dense bodies, like glass, rock-crystal, diamond, &c., it has been argued that there is so exceedingly little of really solid matter even in the densest mass, that *the whole world*, if the atoms could be brought into absolute contact, *might be compressed into a nutshell*. We have as yet no means of determining exactly what relation this idea has to truth.—*Arnott.*

THE WORLD OF ATOMS.

The infinite groups of atoms flying through all time and space, in different directions and under different laws, have interchangeably tried and exhibited every possible mode of rencounter: sometimes repelled from each other by concussion; and sometimes adhering to each other from their own jagged or pointed construction, or from the casual interstices which two or more connected atoms must produce, and which may be just adapted to those of other figures,—as globular, oval, or square. Hence the origin of compound and visible bodies; hence the origin of large masses of matter; hence, eventually, the origin of the world.—*Dr. Good's Book of Nature.*

The great Epicurus speculated on "the plastic nature" of atoms, and attributed to this *nature* the power they possess of arranging themselves into symmetric forms. Modern philosophers satisfy themselves with attraction; and reasoning from analogy, imagine that each atom has a polar system.—*Hunt's Poetry of Science.*

MINUTE ATOMS OF THE ELEMENTS: DIVISIBILITY OF MATTER.

So minute are the parts of the elementary bodies in their ultimate state of division, in which condition they are usually termed *atoms*, as to elude all our powers of inspection, even when aided by the most powerful microscopes. Who can see the particles of gold in a solution of that metal in *aqua regia*, or those of common salt when dissolved in water? Dr. Thomas Thomson has estimated the bulk of an ultimate particle or atom of lead as less than 1/888492000000000th of a cubic inch, and concludes that its weight cannot exceed the 1/310000000000th of a grain.

This curious calculation was made by Dr. Thomson, in order to show to what degree Matter could be divided, and still be sensible to the eye. He dissolved a grain of nitrate of lead in 500,000 grains of water, and passed through the solution a current of sulphuretted hydrogen; when the whole liquid became sensibly discoloured. Now, a grain of water may be regarded as being almost equal to a drop of that liquid, and a drop may be easily spread out so as to cover a square inch of surface. But under an ordinary microscope the millionth of a square inch may be distinguished by the eye. The water, therefore, could be divided into 500,000,000,000 parts. But the lead in a grain of nitrate of lead weighs 0•62 of a grain; an atom of lead, accordingly, cannot weigh more than 1/810000000000th of a grain; while the atom of sulphur, which in combination with the lead rendered it visible, could not weigh more than 1/2015000000000, that is, the two-billionth part of a grain.—*Professor Low; Jameson's Journal*, No. 106.

WEIGHT OF AIR.

Air can be so rarefied that the contents of a cubic foot shall not weigh the tenth part of a grain: if a quantity that would fill a space the hundredth part of an inch in diameter be separated from the rest, the air will still be found there, and we may reasonably conceive that there may be several particles present, though the weight is less than the seventeen-hundred-millionth of a grain.

DURATION OF THE PYRAMID.

The great reason of the duration of the pyramid above all other forms is, that it is most fitted to resist the force of gravitation. Thus the Pyramids of Egypt are the oldest monuments in the world.

INERTIA ILLUSTRATED.

Many things of common occurrence (says Professor Tyndall) are to be explained by reference to the quality of inactivity. We will here state a few of them.

When a railway train is moving, if it strike against any obstacle which arrests its motion, the passengers are thrown forward in the direction in which the train was proceeding. Such accidents often occur on a small scale, in attaching carriages at railway stations. The reason is, that the passengers share the motion of the train, and, as matter, they tend to persist in motion. When the train is suddenly checked, this tendency exhibits itself by the falling forward referred to. In like manner, when a train previously at rest is suddenly set in motion, the tendency of the passengers to remain at rest evinces itself by their falling in a direction opposed to that in which the train moves.

THE LEANING TOWER OF PISA.7

Sir John Leslie used to attribute the stability of this tower to the cohesion of the mortar it is built with being sufficient to maintain it erect, in spite of its being out of the condition required by physics—to wit, that "in order that a column shall stand, a perpendicular let fall from the centre of gravity must fall within the base." Sir John describes the Tower of Pisa to be in violation of this principle; but, according to later authorities, the perpendicular falls within the base.

EARLY PRESENTIMENTS OF CENTRIFUGAL FORCES.

Jacobi, in his researches on the mathematical knowledge of the Greeks, comments on "the profound consideration of nature evinced by Anaxagoras, in whom we read with astonishment a passage asserting that the moon, if the centrifugal force were intermitted, would fall to the earth like a stone from a sling." Anaxagoras likewise applied the same theory of "falling where the force of rotation had been intermitted" to all the material celestial bodies. In Aristotle and Simplicius may also be traced the idea of "the non-falling of heavenly bodies when the rotatory force predominates over the actual falling force, or downward attraction;" and Simplicius mentions that "water in a phial is not spilt when the movement of rotation is more rapid than the downward movement of the water." This is illustrated at the present day by rapidly whirling a pail half-filled with water without spilling a drop.

Plato had a clearer idea than Aristotle of the *attractive force* exercised by the earth's centre on all heavy bodies removed from it; for he was acquainted with the acceleration of falling bodies, although he did not correctly understand the cause. John Philoponus, the Alexandrian, probably in the sixth century, was the first who ascribed the movement of the heavenly bodies to a primitive impulse, connecting with this idea that of the fall of bodies, or the tendency of all substances, whether heavy or light, to reach the ground. The idea conceived by Copernicus, and more clearly expressed by Kepler, who even applied it to the ebb and flow of the ocean, received in 1666 and 1674 a new impulse from Robert Hooke; and next Newton's theory of gravitation presented the grand means of converting the whole of physical astronomy into a true *mechanism of the heavens.*

The law of gravitation knows no exception; it accounts accurately for the most complex motions of the members of our own system; nay more, the paths of double stars, far removed from all appreciable effects of our portion of the universe, are in perfect accordance with its theory.8

HEIGHT OF FALLS.

The fancy of the Greeks delighted itself in wild visions of the height of falls. In Hesiod's *Theogony* it is said, speaking of the fall of the Titans into Tartarus, "if a brazen anvil were to fall from heaven nine days and nine nights long, it would reach the earth on the tenth." This descent of the anvil in 777,600 seconds of time gives an equivalent in distance of 309,424 geographical miles (allowance being made, according to Galle's calculation, for the considerable diminution in force of attraction at planetary distances); therefore 1½ times the distance of the moon from the earth. But, according to the *Iliad*, Hephæstus fell down to Lemnos in one day; "when but a little breath was still in him."—*Note to Humboldt's Cosmos*, vol. iii.

RATE OF THE FALL OF BODIES.

A body falls in gravity precisely 16-1/16 feet in a second, and the velocity increases according to the squares of the time, viz.:

In ¼ (quarter of a second) a body falls	1 foot.
½ (half a second)	4 feet.
1 second	16 ”
2 ditto	64 ”
3 ditto	144 ”

The power of gravity at two miles distance from the earth is four times less than at one mile; at three miles nine times less, and so on. It goes on lessening, but is never destroyed.—*Notes in various Sciences.*

VARIETIES OF SPEED.

A French scientific work states the ordinary rate to be:

per second.

Of a man walking	4 feet.
Of a good horse in harness	12 ”
Of a rein-deer in a sledge on the ice	26 ”
Of an English race-horse	43 ”
Of a hare	88 ”
Of a good sailing ship	19 ”

Of the wind	82	”
Of sound	1038	”
Of a 24-pounder cannon-ball	1300	”

LIFTING HEAVY PERSONS.

One of the most extraordinary pages in Sir David Brewster's *Letters on Natural Magic* is the experiment in which a heavy man is raised with the greatest facility when he is lifted up the instant that his own lungs, and those of the persons who raise him, are inflated with air. Thus the heaviest person in the party lies down upon two chairs, his legs being supported by the one and his back by the other. Four persons, one at each leg, and one at each shoulder, then try to raise him—the person to be raised giving two signals, by clapping his hands. At the first signal, he himself and the four lifters begin to draw a long and full breath; and when the inhalation is completed, or the lungs filled, the second signal is given for raising the person from the chair. To his own surprise, and that of his bearers, he rises with the greatest facility, as if he were no heavier than a feather. Sir David Brewster states that he has seen this inexplicable experiment performed more than once; and he appealed for testimony to Sir Walter Scott, who had repeatedly seen the experiment, and performed the part both of the load and of the bearer. It was first shown in England by Major H., who saw it performed in a large party at Venice, under the direction of an officer of the American navy.9

Sir David Brewster (in a letter to *Notes and Queries*, No. 143) further remarks, that "the inhalation of the lifters the moment the effort is made is doubtless essential, and for this reason: when we make a great effort, either in pulling or lifting, we always fill the chest with air previous to the effort; and when the inhalation is completed, we close the *rima glottidis* to keep the air in the lungs. The chest being thus kept expanded, the pulling or lifting muscles have received as it were a fulcrum round which their power is exerted; and we can thus lift the greatest weight which the muscles are capable of doing. When the chest collapses by the escape of the air, the lifters lose their muscular power; reinhalation of air by the liftee can certainly add nothing to the power of the lifters, or diminish his own weight, which is only increased by the weight of the air which he inhales."

"FORCE CAN NEITHER BE CREATED NOR DESTROYED."

Professor Faraday, in his able inquiry upon "the Conservation of Force," maintains that to admit that force may be destructible, or can altogether disappear, would be to admit that matter could be uncreated; for we know matter only by its forces. From his many illustrations we select the following:

The indestructibility of individual matter is a most important case of the Conservation of Chemical Force. A molecule has been endowed with powers which give rise in it to various qualities; and those never change, either in their nature or amount. A particle of oxygen is ever a particle of oxygen; nothing can in the least wear it. If it enters into combination, and disappears as oxygen; if it pass through a thousand combinations—animal, vegetable, mineral; if it lie hid for a thousand years, and then be evolved,—it is oxygen with the first qualities, neither more nor less. It has all its original force, and only that; the amount of force which it disengaged when hiding itself, has again to be employed in a reverse direction when it is set at liberty: and if, hereafter, we should decompose oxygen, and find it compounded of other particles, we should only increase the strength of the proof of the conservation of force; for we should have a right to say of these particles, long as they have been hidden, all that we could say of the oxygen itself.

In conclusion, he adds:

Let us not admit the destruction or creation of force without clear and constant proof. Just as the chemist owes all the perfection of his science to his dependence on the certainty of gravitation applied by the balance, so may the physical philosopher expect to find the greatest security and the utmost aid in the principle of the conservation of force. All that we have that is good and safe—as the steam-engine, the electric telegraph, &c.—witness to that principle; it would require a perpetual motion, a fire without heat, heat without a source, action without reaction, cause without effect, or effect without cause, to displace it from its rank as a law of nature.

NOTHING LOST IN THE MATERIAL WORLD.

"It is remarkable," says Kobell in his *Mineral Kingdom*, "how a change of place, a circulation as it were, is appointed for the inanimate or naturally immovable things upon the earth; and how new conditions, new creations, are continually developing themselves in this way. I will not enter here into the evaporation of water, for instance from the widely-spreading ocean; how the clouds produced by this pass over into foreign lands and then fall again to the earth as rain, and how this wandering water is, partly at least, carried along new journeys, returning after various voyages to its original home: the mere mechanical phenomena, such as the transfer of seeds by the winds or by birds, or the decomposition of the surface of the earth by the friction of the elements, suffice to illustrate this."

TIME AN ELEMENT OF FORCE.

Professor Faraday observes that Time is growing up daily into importance as an element in the exercise of Force, which he thus strikingly illustrates:

The earth moves in its orbit of time; the crust of the earth moves in time; light moves in time; an electro-magnet requires time for its charge by an electric current: to inquire, therefore, whether power, acting either at sensible or insensible distances, always acts in *time*, is not to be metaphysical; if it acts in time and across space, it must act by physical lines of force; and our view of the nature of force may be affected to the extremest degree by the conclusions which experiment and observation on time may supply, being perhaps finally determinable only by them. To inquire after the possible time in which gravitating, magnetic, or electric force is exerted, is no more metaphysical than to mark the times of the hands of a clock in their progress; or that of the temple of Serapis, and its ascents and descents; or the periods of the occultation of Jupiter's satellites; or that in which the light comes from them to the earth. Again, in some of the known cases of the action of time something happens while *the time* is passing which did not happen before, and does not continue after; it is therefore not metaphysical to expect an effect in *every* case, or to endeavour to discover its existence and determine its nature.

CALCULATION OF HEIGHTS AND DISTANCES.

By the assistance of a seconds watch the following interesting calculations may be made:

If a traveller, when on a precipice or on the top of a building, wish to ascertain the height, he should drop a stone, or any other substance sufficiently heavy not to be impeded by the resistance of the atmosphere; and the number of seconds which elapse before it reaches the bottom, carefully noted on a seconds watch, will give the height. For the stone will fall through the space of 16-1/8 feet during the first second, and will increase in rapidity as the square of the time employed in the fall: if, therefore, 16-1/8 be multiplied by the number of seconds the stone has taken to fall, this product also multiplied by the same number of seconds will give the height. Suppose the stone takes five seconds to reach the bottom:

$$16\text{-}1/8 \times 5 = 80\text{-}5/8 \times 5 = 403\text{-}1/8\text{, height of the precipice.}$$

The Count Xavier de Maistre, in his *Expédition nocturne autour de ma Chambre*, anxious to ascertain the exact height of his room from the ground on which Turin is built, tells us he proceeded as follows: "My heart beat quickly, and I just counted three pulsations from the instant I dropped my slipper until I heard the sound as it fell in the street, which, according to the calculations made of the time taken by bodies in their accelerated fall, and of that employed by the sonorous undulations of the air to arrive from the street to my ear, gave the height of my apartment as 94 feet 3 inches 1 tenth (French measure), supposing that my heart, agitated as it was, beat 120 times in a minute."

A person travelling may ascertain his rate of walking by the aid of a slight string with a piece of lead at one end, and the use of a seconds watch; the string being knotted at distances of 44 feet, the 120th part of an English mile, and bearing the same proportion to a mile that half a minute bears to an hour. If the traveller, when going at his usual rate, drops the lead, and suffers the string to slip through his hand, the number of knots which pass in half a minute indicate the number of miles he walks in an hour. This contrivance is similar to a *log-line* for ascertaining a ship's rate at sea: the lead is enclosed in wood (whence the name *log*), that it may float, and the divisions, which are called *knots*, are measured for nautical miles. Thus, if ten knots are passed in half a minute, they show that the vessel is sailing at the rate of ten knots, or miles, an hour: a seconds watch would here be of great service, but the half-minute sand-glass is in general use.

The rapidity of a river may be ascertained by throwing in a light floating substance, which, if not agitated by the wind, will move with the same celerity as the water: the distance it floats in a certain number of seconds will give the rapidity of the stream; and this indicates the height of its source, the nature of its bottom, &c.—See *Sir Howard Douglas on Bridges. Thomson's Time and Time-keepers.*

SAND IN THE HOUR-GLASS.

It is a noteworthy fact, that the flow of Sand in the Hour-glass is perfectly equable, whatever may be the quantity in the glass; that is, the sand runs no faster when the upper half of the glass is quite full than when it is nearly empty. It would, however, be natural enough to conclude, that when full of sand it would be more swiftly urged through the aperture than when the glass was only a quarter full, and near the close of the hour.

The fact of the even flow of sand may be proved by a very simple experiment. Provide some silver sand, dry it over or before the fire, and pass it through a tolerably fine sieve. Then take a tube, of any length or diameter, closed at one end, in which make a small hole, say the eighth of an inch; stop this with a peg, and fill up the tube with the sifted sand. Hold the tube steadily, or fix it to a wall or frame at any height from a table; remove the peg, and permit the sand to flow in any measure for any given time, and note the quantity. Then let the tube be emptied, and only half or a quarter filled with sand; measure again for a like time, and the same quantity of sand will flow: even if you press the sand in the tube with a ruler or stick, the flow of the sand through the hole will not be increased.

The above is explained by the fact, that when the sand is poured into the tube, it fills it with a succession of conical heaps; and that all the weight which the bottom of the tube sustains is only that of the heap which *first* falls upon it, as the succeeding heaps do not press downward, but only against the sides or walls of the tube.

FIGURE OF THE EARTH.

By means of a purely astronomical determination, based upon the action which the earth exerts on the motion of the moon, or, in other words, on the inequalities in lunar longitudes and latitudes, Laplace has shown in one single result the mean Figure of the Earth.

It is very remarkable that an astronomer, without leaving his observatory, may, merely by comparing his observations with mean analytical results, not only be enabled to determine with exactness the size and degree of ellipticity of the earth, but also its distance from the sun and moon; results that otherwise could only be arrived at by long and arduous expeditions to the most remote parts of both hemispheres. The moon may therefore, by the observation of its movements, render appreciable to the higher departments of astronomy the ellipticity of the earth, as it taught the early astronomers the rotundity of our earth by means of its eclipses.—*Laplace's Expos. du Syst. du Monde.*

HOW TO ASCERTAIN THE EARTH'S MAGNITUDE.

Sir John Herschel gives the following means of approximation. It appears by observation that two points, each ten feet above the surface, cease to be visible from each other over still water, and, in average atmospheric circumstances, at a distance of about eight miles. But 10 feet is the 528th part of a mile; so that half their distance, or four miles, is to the height of each as 4×528, or 2112:1, and therefore in the same proportion to four miles is the length of the earth's diameter. It must, therefore, be equal to $4 \times 2112 = 8448$, or in round numbers, about 8000 miles, which is not very far from the truth.

The excess is, however, about 100 miles, or 1/80th part. As convenient numbers to remember, the reader may bear in mind, that in our latitude there are just as many thousands of feet in a degree of the meridian as there are days in the year

(365); that, speaking loosely, a degree is about seventy British statute miles, and a second about 100 feet; that the equatorial circumference of the earth is a little less than 25,000 miles (24,899), and the ellipticity or polar flattening amounts to 1/300th part of the diameter.—*Outlines of Astronomy.*

MASS AND DENSITY OF THE EARTH.

With regard to the determination of the Mass and Density of the Earth by direct experiment, we have, in addition to the deviations of the pendulum produced by mountain masses, the variation of the same instruments when placed in a mine 1200 feet in depth. The most recent experiments were conducted by Professor Airy, in the Harton coal-pit, near South Shields:10 the oscillations of the pendulum at the bottom of the pit were compared with those of a clock above; the beats of the clock were transferred below for comparison by an electrio wire; and it was thus determined that a pendulum vibrating seconds at the mouth of the pit would gain 2¼ seconds per day at its bottom. The final result of the calculations depending on this experiment, which were published in the*Philosophical Transactions* of 1856, gives 6•565 for the mean density of the earth. The celebrated Cavendish experiment, by means of which the density of the earth was determined by observing the attraction of leaden balls on each other, has been repeated in a manner exhibiting an astonishing amount of skill and patience by the late Mr. F. Baily.11 The result of these experiments, combined with those previously made, gives as a mean result 5•441 as the earth's density, when compared with water; thus confirming one of Newton's astonishing divinations, that the mean density of the earth would be found to be between five and six times that of water.

Humboldt is, however, of opinion that "we know only the mass of the whole earth and its mean density by comparing it with the open strata, which alone are accessible to us. In the interior of the earth, where all knowledge of its chemical and mineralogical character fails, we are limited to as pure conjecture as in the remotest bodies that revolve round the sun. We can determine nothing with certainty regarding the depth at which the geological strata must be supposed to be in a state of softening or of liquid fusion, of the condition of fluids when heated under an enormous pressure, or of the law of the increase of density from the upper surface to the centre of the earth."—*Cosmos,* vol. i.

In M. Foucault's beautiful experiment, by means of the vibration of a long pendulum, consisting of a heavy mass of metal suspended by a long wire from a strong fixed support, is demonstrated to the eye the rotation of the earth. The Gyroscope of the same philosopher is regarded not as a mere philosophical toy; but the principles of dynamics, by means of which it is made to demonstrate the earth's rotation on its own axis, are explained with the greatest clearness. Thus the ingenuity of M. Foucault, combined with a profound knowledge of mechanics, has obtained proofs of one of the most interesting problems of astronomy from an unsuspected source.

THE EARTH AND MAN COMPARED.

The Earth—speaking roundly—is 8000 miles in diameter; the atmosphere is calculated to be fifty miles in altitude; the loftiest mountain peak is estimated at five miles above the level of the sea, for this height has never been visited by man; the deepest mine that he has formed is 1650 feet; and his own stature does not average six feet. Therefore, if it were possible for him to construct a globe 800 feet—or twice the height of St. Paul's Cathedral—in diameter, and to place upon any one point of its surface an atom of 1/4380th of an inch in diameter, and 1/720th of an inch in height, it would correctly denote the proportion that man bears to the earth upon which he moves.

When by measurements, in which the evidence of the method advances equally with the precision of the results, the volume of the earth is reduced to the millionth part of the volume of the sun; when the sun himself, transported to the region of the stars, takes up a very modest place among the thousands of millions of those bodies that the telescope has revealed to us; when the 38,000,000 of leagues which separate the earth from the sun have become, by reason of their comparative smallness, a base totally insufficient for ascertaining the dimensions of the visible universe; when even the swiftness of the luminous rays (77,000 leagues per second) barely suffices for the common valuations of science; when, in short, by a chain of irresistible proofs, certain stars have retired to distances that light could not traverse in less than a million of years;—we feel as if annihilated by such immensities. In assigning to man and to the planet that he inhabits so small a position in the material world, astronomy seems really to have made progress only to humble us.—*Arago.*

MEAN TEMPERATURE OF THE EARTH'S SURFACE.

Professor Dove has shown, by taking at all seasons the mean of the temperature of points diametrically opposite to each other, that the mean temperature *of the whole earth's surface* in June considerably exceeds that in December. This result, which is at variance with the greater proximity of the sun in December, is, however, due to a totally different and very powerful cause,—the greater amount of land in that hemisphere which has its summer solstice in June (*i. e.* the northern); and the fact is so explained by him. The effect of land under sunshine is to throw heat into the general atmosphere, and to distribute it by the carrying power of the latter over the whole earth. Water is much less effective in this respect, the heat penetrating its depths and being there absorbed; so that the surface never acquires a very elevated temperature, even under the equator.—*Sir John Herschel's Outlines.*

TEMPERATURE OF THE EARTH STATIONARY.

Although, according to Bessel, 25,000 cubic miles of water flow in every six hours from one quarter of the earth to another, and the temperature is augmented by the ebb and flow of every tide, all that we know with certainty is, that the *resultant effect* of all the thermal agencies to which the earth is exposed has undergone no perceptible change within the historic period. We owe this fine deduction to Arago. In order that the *date palm* should ripen its fruit, the mean temperature of the place must exceed 70 deg. Fahr.; and, on the other hand, the *vine* cannot be cultivated successfully when the temperature is 72 deg. or upwards. Hence the mean temperature of any place at which these two plants flourished and bore fruit must lie between these narrow limits, *i. e.* could not differ from 71 deg. Fahr. by more than a single degree. Now from the Bible we learn that both plants were *simultaneously* cultivated in the central valleys of Palestine in the time of Moses; and its then temperature is thus definitively determined. It is the same at the present time; so that the mean temperature of this portion of the globe has not sensibly altered in the course of thirty-three centuries.

THEORY OF CRYSTALLISATION.

Professor Plücker has ascertained that certain crystals, in particular the cyanite, "point very well to the north by the magnetic power of the earth only. It is a true compass-needle; and more than that, you may obtain its declination." Upon this Mr. Hunt remarks: "We must remember that this crystal, the cyanite, is a compound of silica and alumina only. This is the amount of experimental evidence which science has afforded in explanation of the conditions under which nature pursues her wondrous work of crystal formation. We see just sufficient of the operation to be convinced that the luminous star which shines in the brightness of heaven, and the cavern-secreted gem, are equally the result of forces which are known to us in only a few of their modifications."—*Poetry of Science.*

Gay Lussac first made the remark, that a crystal of potash-alum, transferred to a solution of ammonia-alum, continued to increase without its form being modified, and might thus be covered with alternate layers of the two alums, preserving its regularity and proper crystalline figure. M. Beudant afterwards observed that other bodies, such as the sulphates of iron and copper, might present themselves in crystals of the same form and angles, although the form was not a simple one, like that of alum. But M. Mitscherlich first recognised this correspondence in a sufficient number of cases to prove that it was a general consequence of similarity of composition in different bodies.—*Graham's Elements of Chemistry.*

IMMENSE CRYSTALS.

Crystals are found in the most microscopic character, and of an exceedingly large size. A crystal of quartz at Milan is three feet and a quarter long, and five feet and a half in circumference: its weight is 870 pounds. Beryls have been found in New Hampshire measuring four feet in length.—*Dana.*

VISIBLE CRYSTALLISATION.

Professor Tyndall, in a lecture delivered by him at the Royal Institution, London, on the properties of Ice, gave the following interesting illustration of crystalline force. By perfectly cleaning a piece of glass, and placing on it a film of a solution of chloride of ammonium or sal ammoniac, the action of crystallisation was shown to the whole audience. The glass slide was placed in a microscope, and the electric light passing through it was concentrated on a white disc. The image of the crystals, as they started into existence, and shot across the disc in exquisite arborescent and symmetrical forms, excited the admiration of every one. The lecturer explained that the heat, causing the film of moisture to evaporate, brought the particles of salt sufficiently near to exercise the crystalline force, the result being the beautiful structure built up with such marvellous rapidity.

UNION OF MINERALOGY AND GEOMETRY.

It is a peculiar characteristic of minerals, that while plants and animals differ in various regions of the earth, mineral matter of the same character may be discovered in any part of the world,—at the Equator or towards the Poles; at the summit of the loftiest mountains, and in works far beneath the level of the sea. The granite of Australia does not necessarily differ from that of the British islands; and ores of the same metals (the proper geological conditions prevailing) may be found of the same general character in all regions. Climate and geographical position have no influence on the composition of mineral substances.

This uniformity may, in some measure, have induced philosophers to seek its extension to the forms of crystallography. About 1760 (says Mr. Buckle, in his *History of Civilization*), Romé de Lisle set the first example of studying crystals, according to a scheme so large as to include all the varieties of their primary forms, and to account for their irregularities and the apparent caprice with which they were arranged. In this investigation he was guided by the fundamental assumption, that what is called an irregularity is in truth perfectly regular, and that the operations of nature are invariable. Haüy applied this great idea to the almost innumerable forms in which minerals crystallise. He thus achieved a complete union between mineralogy and geometry; and, bringing the laws of space to bear on the molecular arrangements of matter, he was able to penetrate into the intimate structure of crystals. By this means he proved that the secondary forms of all crystals are derived from their primary forms by a regular process of decrement; and that when a substance is passing from a liquid to a solid state, its particles cohere, according to a scheme which provides for every possible change, since it includes even those subsequent layers which alter the ordinary type of the crystal, by disturbing its natural symmetry. To ascertain that such violations of symmetry are susceptible of mathematical calculation, was to make a vast addition to our knowledge; and, by proving that even the most uncouth and singular forms are the natural results of their antecedents, Haüy laid the foundation of what may be called the pathology of the inorganic world. However paradoxical such a notion may appear, it is certain that symmetry is to crystals what health is to animals; so that an irregularity of shape in the first corresponds with an appearance of disease in the second.—See *Hist. Civilization*, vol. i.

REPRODUCTIVE CRYSTALLISATION.

The general belief that only organic beings have the power of reproducing lost parts has been disproved by the experiments of Jordan on crystals. An octohedral crystal of alum was fractured; it was then replaced in a solution, and after a few days its injury was seen to be repaired. The whole crystal had of course increased in size; but the increase on the broken surface had been so much greater that a perfect octohedral form was regained.—*G. H. Lewes.*

This remarkable power possessed by crystals, in common with animals, of repairing their own injuries had, however, been thus previously referred to by Paget, in his *Pathology*, confirming the experiments of Jordan on this curious subject: "The ability to repair the damages sustained by injury ... is not an exclusive property of living beings; for even crystals will repair themselves when, after pieces have been broken from them, they are placed in the same conditions in which they were first formed."

GLASS BROKEN BY SAND.

In some glass-houses the workmen show glass which has been cooled in the open air; on this they let fall leaden bullets without breaking the glass. They afterwards desire you to let a few grains of sand fall upon the glass, by which it is broken into a thousand pieces. The reason of this is, that the lead does not scratch the surface of the glass; whereas the sand, being sharp and angular, scratches it sufficiently to produce the above effect.

Sound and Light.

SOUNDING SAND.

Mr. Hugh Miller, the geologist, when in the island of Eigg, in the Hebrides, observed that a musical sound was produced when he walked over the white dry sand of the beach. At each step the sand was driven from his footprint, and the noise was simultaneous with the scattering of the sand; the cause being either the accumulated vibrations of the air when struck by the driven sand, or the accumulated sounds occasioned by the mutual impact of the particles of sand against each other. If a musket-ball passing through the air emits a whistling note, each individual particle of sand must do the same, however faint be the note which it yields; and the accumulation of these infinitesimal vibrations must constitute an audible sound, varying with the number and velocity of the moving particles. In like manner, if two plates of silex or quartz, which are but crystals of sand, give out a musical sound when mutually struck, the impact or collision of two minute crystals or particles of sand must do the same, in however inferior a degree; and the union of all these sounds, though singly imperceptible, may constitute the musical notes of "the Mountain of the Bell" in Arabia Petræa, or the lesser sounds of the trodden sea-beach of Eigg.—*North-British Review*, No. 5.

INTENSITY OF SOUND IN RAREFIED AIR.

The experiences during ascents of the highest mountains are contradictory. Saussure describes the sounds on the top of Mont Blanc as remarkably weak: a pistol-shot made no more noise than an ordinary Chinese cracker, and the popping of a bottle of champagne was scarcely audible. Yet Martius, in the same situation, was able to distinguish the voices of the guides at a distance of 1340 feet, and to hear the tapping of a lead pencil upon a metallic surface at a distance of from 75 to 100 feet.

MM Wertheim and Breguet have propagated sound over the wire of an electric telegraph at the rate of 11,454 feet per second.

DISTANCE AT WHICH THE HUMAN VOICE MAY BE HEARD.

Experience shows that the human voice, under favourable circumstances, is capable of filling a larger space than was ever probably enclosed within the walls of a single room. Lieutenant Foster, on Parry's third Arctic expedition, found that he could converse with a man across the harbour of Port Bowen, a distance of 6696 feet, or about one mile and a quarter. Dr. Young records that at Gibraltar the human voice has been heard at a distance of ten miles. If sound be prevented from spreading and losing itself in the air, either by a pipe or an extensive flat surface, as a wall or still water, it may be conveyed to a great distance. Biot heard a flute clearly through a tube of cast-iron (the water-pipes of Paris) 3120 feet long: the lowest whisper was distinctly heard; indeed, the only way not to be heard was not to speak at all.

THE ROAR OF NIAGARA.

The very nature of the sound of running water pronounces its origin to be the bursting of bubbles: the impact of water against water is a comparatively subordinate cause, and could never of itself occasion the murmur of a brook; whereas, in streams which Dr. Tyndall has examined, he, in all cases where a ripple was heard, discovered bubbles caused by the broken column of water. Now, were Niagara continuous, and without lateral vibration, it would be as silent as a cataract of ice. In all probability, it has its "contracted sections," after passing which it is broken into detached masses, which, plunging successively upon the air-bladders formed by their precursors, suddenly liberate their contents, and thus create *the thunder of the waterfall.*

FIGURES PRODUCED BY SOUND.

Stretch a sheet of wet paper over the mouth of a glass tumbler which has a footstalk, and glue or paste the paper at the edges. When the paper is dry, strew dry sand thinly upon its surface. Place the tumbler on a table, and hold immediately above it, and parallel to the paper, a plate of glass, which you also strew with sand, having previously rubbed the edges smooth with emery powder. Draw a violin-bow along any part of the edges; and as the sand upon the glass is made to vibrate, it will form various figures, which will be accurately imitated by the sand upon the paper; or if a violin or flute be played within a few inches of the paper, they will cause the sand upon its surface to form regular lines and figures.

THE TUNING-FORK A FLUTE-PLAYER.

Take a common tuning-fork, and on one of its branches fasten with sealing-wax a circular piece of card of the size of a small wafer, or sufficient nearly to cover the aperture of a pipe, as the sliding of the upper end of a flute with the mouth stopped: it may be tuned in unison with the loaded tuning-fork by means of the movable stopper or card, or the fork may be loaded till the unison is perfect. Then set the fork in vibration by a blow on the unloaded branch, and hold the card closely over the mouth of the pipe, as in the engraving, when a note of surprising clearness and strength will be heard. Indeed a flute may be made to "speak" perfectly well, by holding close to the opening a vibrating tuning-fork, while the fingering proper to the note of the fork is at the same time performed.

THEORY OF THE JEW'S HARP.

If you cause the tongue of this little instrument to vibrate, it will produce a very low sound; but if you place it before a cavity (as the mouth) containing a column of air, which vibrates much faster, but in the proportion of any simple multiple, it will then produce other higher sounds, dependent upon the reciprocation of that portion of the air. Now the bulk of air in the mouth can be altered in its form, size, and other circumstances, so as to produce by reciprocation many different sounds; and these are the sounds belonging to the Jew's Harp.

A proof of this fact has been given by Mr. Eulenstein, who fitted into a long metallic tube a piston, which being moved, could be made to lengthen or shorten the efficient column of air within at pleasure. A Jew's Harp was then so fixed that it could be made to vibrate before the mouth of the tube, and it was found that the column of air produced a series of sounds, according as it was lengthened or shortened; a sound being produced whenever the length of the column was such that its vibrations were a multiple of those of the Jew's Harp.

SOLAR AND ARTIFICIAL LIGHT COMPARED.

The most intensely ignited solid (produced by the flame of Lieutenant Drummond's oxy-hydrogen lamp directed against a surface of chalk) appears only as black spots on the disc of the sun, when held between it and the eye; or in other words, Drummond's light is to the light of the sun's disc as 1 to 146. Hence we are doubly struck by the felicity with which Galileo, as early as 1612, by a series of conclusions on the smallness of the distance from the sun at which the disc of Venus was no longer visible to the naked eye, arrived at the result that the blackest nucleus of the sun's spots was more luminous than the brightest portions of the full moon. (See "The Sun's Light compared with Terrestrial Lights," in *Things not generally Known*, pp. 4, 5.)

SOURCE OF LIGHT.

Mr. Robert Hunt, in a lecture delivered by him at the Russell Institution, "On the Physics of a Sunbeam," mentions some experiments by Lord Brougham on the sunbeam, in which, by placing the edge of a sharp knife just within the limit of the light, the ray was inflected from its previous direction, and coloured red; and when another knife was placed on the opposite side, it was deflected, and the colour was blue. These experiments (says Mr. Hunt) seem to confirm Sir Isaac Newton's theory, that light is a fluid emitted from the sun.

THE UNDULATORY SCALE OF LIGHT.

The white light of the sun is well known to be composed of several coloured rays; or rather, according to the theory of undulations, when the rate at which a ray vibrates is altered, a different sensation is produced upon the optic nerve. The analytical examination of this question shows that to produce a red colour the ray of light must give 37,640 undulations in an inch, and 458,000,000,000,000 in a second. Yellow light requires 44,000 undulations in an inch, and 535,000,000,000,000 in a second; whilst the effect of blue results from 51,110 undulations within an inch, and 622,000,000,000,000 of waves in a second of time.—*Hunt's Poetry of Science.*

VISIBILITY OF OBJECTS.

In terrestrial objects, the form, no less than the modes of illumination, determines the magnitude of the smallest angle of vision for the naked eye. Adams very correctly observed that a long and slender staff can be seen at a much greater distance than a square whose sides are equal to the diameter of the staff. A stripe may be distinguished at a greater distance than a spot, even when both are of the same diameter.

The *minimum* optical visual angle at which terrestrial objects can be recognised by the naked eye has been gradually estimated lower and lower, from the time when Robert Hooke fixed it exactly at a full minute, and Tobias Meyer required 34″ to perceive a black speck on white paper, to the period of Leuwenhoeck's experiments with spiders' threads, which are visible to ordinary sight at an angle of 4″•7. In Hueck's most accurate experiments on the problem of the movement of the crystalline lens, white lines on a black ground were seen at an angle of 1″•2; a spider's thread at 0″•6; and a fine glistening wire at scarcely 0″•2.

Humboldt, when at Chillo, near Quito, where the crests of the volcano of Pichincha lay at a horizontal distance of 90,000 feet, was much struck by the circumstance that the Indians standing near distinguished the figure of Bonpland (then on an expedition to the volcano), as a white point moving on the black basaltic sides of the rock, sooner than Humboldt could discover him with a telescope. Bonpland was enveloped in a white cotton poncho: assuming the breadth across the shoulders to vary from three to five feet, according as the mantle clung to the figure or fluttered in the breeze, and judging from the known distance, the angle at which the moving object could be distinctly seen varied from 7″ to 12″. White objects on a black ground are, according to Hueck, distinguished at a greater distance than black objects on a white ground.

Gauss's heliotrope light has been seen with the naked eye reflected from the Brocken on Hohenhagen at a distance of about 227,000 feet, or more than 42 miles; being frequently visible at points in which the apparent breadth of a three-inch mirror was only 0″•43.

THE SMALLEST BRIGHT BODIES.

Ehrenberg has found from experiments on the dust of diamonds, that a diamond superficies of 1/100th of a line in diameter presents a much more vivid light to the naked eye than one of quicksilver of the same diameter. On pressing small globules of quicksilver on a glass micrometer, he easily obtained smaller globules of the 1/100th to the 1/2000th of a line in diameter. In the sunshine he could only discern the reflection of light, and the existence of such globules as were 1/300th of a line in diameter, with the naked eye. Smaller ones did not affect his eye; but he remarked that the actual bright part of the globule did not amount to more than 1/900th of a line in diameter. Spider threads of 1/2000th in diameter were still discernible from their lustre. Ehrenberg concludes that there are in organic bodies magnitudes capable of direct proof which are in diameter 1/100000 of a line; and others, that can be indirectly proved, which may be less than a six-millionth part of a Parisian line in diameter.

VELOCITY OF LIGHT.

It is scarcely possible so to strain the imagination as to conceive the Velocity with which Light travels. "What mere assertion will make any man believe," asks Sir John Herschel, "that in one second of time, in one beat of the pendulum of a clock, a ray of light travels over 192,000 miles; and would therefore perform the tour of the world in about the same time that it requires to wink with our eyelids, and in much less time than a swift runner occupies in taking a single stride?" Were a cannon-ball shot directly towards the sun, and were it to maintain its full speed, it would be twenty years in reaching it; and yet light travels through this space in seven or eight minutes.

The result given in the *Annuaire* for 1842 for the velocity of light in a second is 77,000 leagues, which corresponds to 215,834 miles; while that obtained at the Pulkowa Observatory is 189,746 miles. William Richardson gives as the result of the passage of light from the sun to the earth 8′ 19″•28, from which we obtain a velocity of 215,392 miles in a second.—*Memoirs of the Astronomical Society*, vol. iv.

In other words, light travels a distance equal to eight times the circumference of the earth between two beats of a clock. This is a prodigious velocity; but the measure of it is very certain.—*Professor Airy.*

The navigator who has measured the earth's circuit by his hourly progress, or the astronomer who has paced a degree of the meridian, can alone form a clear idea of velocity, when we tell him that light moves through a space equal to the circumference of the earth in *the eighth part of a second*—in the twinkling of an eye.

Could an observer, placed in the centre of the earth, see this moving light, as it describes the earth's circumference, it would appear a luminous ring; that is, the impression of the light at the commencement of its journey would continue on the retina till the light had completed its circuit. Nay, since the impression of light continues longer than the *fourth* part of a second, *two* luminous rings would be seen, provided the light made *two* rounds of the earth, and in paths not coincident.

APPARATUS FOR THE MEASUREMENT OF LIGHT.

Humboldt enumerates the following different methods adopted for the Measurement of Light: a comparison of the shadows of artificial lights, differing in numbers and distance; diaphragms; plane-glasses of different thickness and colour; artificial stars formed by reflection on glass spheres; the juxtaposition of two seven-feet telescopes, separated by a distance which the observer could pass in about a second; reflecting instruments in which two stars can be simultaneously seen and compared, when the telescope has been so adjusted that the star gives two images of like intensity; an apparatus having (in front of the object-glass) a mirror and diaphragms, whose rotation is measured on a ring; telescopes with divided object-glasses, on either half of which the stellar light is received through a prism; astrometers, in which a prism reflects the image of the moon or Jupiter, and concentrates it through a lens at different distances into a star more or less bright.—*Cosmos*, vol. iii.

HOW FIZEAU MEASURED THE VELOCITY OF LIGHT.

This distinguished physicist has submitted the Velocity of Light to terrestrial measurement by means of an ingeniously constructed apparatus, in which artificial light (resembling stellar light), generated from oxygen and hydrogen, is made to pass back, by means of a mirror, over a distance of 28,321 feet to the same point from which it emanated. A disc, having 720 teeth, which made 12•6 rotations in a second, alternately obscured the ray of light and allowed it to be seen between the teeth on the margin. It was supposed, from the marking of a counter, that the artificial light traversed 56,642 feet, or the distance to and from the stations, in 1/1800th part of a second, whence we obtain a velocity of 191,460 miles in a second.[12] This result approximates most closely to Delambre's (which was 189,173 miles), as obtained from Jupiter's satellites.

The invention of the rotating mirror is due to Wheatstone, who made an experiment with it to determine the velocity of the propagation of the discharge of a Leyden battery. The most striking application of the idea was made by Fizeau and Foucault, in 1853, in carrying out a proposition made by Arago, soon after the invention of the mirror: we have here determined in a distance of twelve feet no less than the velocity with which light is propagated, which is known to be nearly 200,000 miles a second; the distance mentioned corresponds therefore to the 77-millionth part of a second. The object of these measurements was to compare the velocity of light in air with its velocity in water; which, when the length is greater, is not sufficiently transparent. The most complete optical and mechanical aids are here necessary: the mirror of Foucault made from 600 to 800 revolutions in a second, while that of Fizeau performed 1200 to 1500 in the same time.— *Prof. Helmholtz on the Methods of Measuring very small Portions of Time.*

WHAT IS DONE BY POLARISATION OF LIGHT.

Malus, in 1808, was led by a casual observation of the light of the setting sun, reflected from the windows of the Palais de Luxembourg, at Paris, to investigate more thoroughly the phenomena of double refraction, of ordinary and of chromatic polarisation, of interference and of diffraction of light. Among his results may be reckoned the means of distinguishing between direct and reflected light; the power of penetrating, as it were, into the constitution of the body of the sun and of its luminous envelopes; of measuring the pressure of atmospheric strata, and even the smallest amount of water they contain; of ascertaining the depths of the ocean and its rocks by means of a tourmaline plate; and in accordance with Newton's prediction, of comparing the chemical composition of several substances with their optical effects.

Arago, in a letter to Humboldt, states that by the aid of his polariscope, he discovered, before 1820, that the light of all terrestrial objects in a state of incandescence, whether they be solid or liquid, is natural, so long as it emanates from the object in perpendicular rays. On the other hand, if such light emanate at an acute angle, it presents manifest proofs of polarisation. This led M. Arago to the remarkable conclusion, that light is not generated on the surface of bodies only, but that some portion is actually engendered within the substance itself, even in the case of platinum.

A ray of light which reaches our eyes after traversing many millions of miles, from, the remotest regions of heaven, announces, as it were of itself, in the polariscope, whether it is reflected or refracted, whether it emanates from a solid or fluid or gaseous body; it announces even the degree of its intensity.—*Humboldt's Cosmos*, vols. i. and ii.

MINUTENESS OF LIGHT.

There is something wonderful, says Arago, in the experiments which have led natural philosophers legitimately to talk of the different sides of a ray of light; and to show that millions and millions of these rays can simultaneously pass through the eye of a needle without interfering with each other!

THE IMPORTANCE OF LIGHT.

Light affects the respiration of animals just as it affects the respiration of plants. This is novel doctrine, but it is demonstrable. In the day-time we expire more carbonic acid than during the night; a fact known to physiologists, who explain it as the effect of sleep: but the difference is mainly owing to the presence or absence of sunlight; for sleep, as sleep, *increases*, instead of diminishing, the amount of carbonic acid expired, and a man sleeping will expire more carbonic acid than if he lies quietly awake under the same conditions of light and temperature; so that if less is expired during the night than during the day, the reason cannot be sleep, but the absence of light. Now we understand why men are sickly and stunted who live in narrow streets, alleys, and cellars, compared with those who, under similar conditions of poverty and dirt, live in the sunlight.—*Blackwood's Edinburgh Magazine*, 1858.

The influence of light on the colours of organised creation is well shown in the sea. Near the shores we find seaweeds of the most beautiful hues, particularly on the rocks which are left dry by the tides; and the rich tints of the actiniæ which inhabit shallow water must often have been observed. The fishes which swim near the surface are also distinguished by the variety of their colours, whereas those which live at greater depths are gray, brown, or black. It has been found that after a certain depth, where the quantity of light is so reduced that a mere twilight prevails, the inhabitants of the ocean become nearly colourless.—*Hunt's Poetry of Science.*

ACTION OF LIGHT ON MUSCULAR FIBRES.

That light is capable of acting on muscular fibres, independently of the influence of the nerves, was mentioned by several of the old anatomists, but repudiated by later authorities. M. Brown Séquard has, however, proved to the Royal Society that some portions of muscular fibre—the iris of the eye, for example—are affected by light independently of any reflex action of the nerves, thereby confirming former experiences. The effect is produced by the illuminating rays only, the chemical and heat rays remaining neutral. And not least remarkable is the fact, that the iris of an eel showed itself susceptible of the excitement *sixteen days after the eyes were removed from the creature's head.* So far as is yet known, this muscle is the only one on which light thus takes effect.—*Phil. Trans. 1857.*

LIGHT NIGHTS.

It is not possible, as well-attested facts prove, perfectly to explain the operations at work in the much-contested upper boundaries of our atmosphere. The extraordinary lightness of whole nights in the year 1831, during which small print might be read at midnight in the latitudes of Italy and the north of Germany, is a fact directly at variance with all that we know, according to the most recent and acute researches on the crepuscular theory and the height of the atmosphere.—*Biot.*

PHOSPHORESCENCE OF PLANTS.

Mr. Hunt recounts these striking instances. The leaves of the *œnothera macrocarpa* are said to exhibit phosphoric light when the air is highly charged with electricity. The agarics of the olive-grounds of Montpelier too have been observed to be luminous at night; but they are said to exhibit no light, even in darkness, *during the day.* The subterranean passages of the coal-mines near Dresden are illuminated by the phosphorescent light of the *rhizomorpha phosphoreus,* a peculiar fungus. On the leaves of the Pindoba palm grows a species of agaric which is exceedingly luminous at night; and many varieties of the lichens, creeping along the roofs of caverns, lend to them an air of enchantment by the soft and clear light which they diffuse. In a small cave near Penryn, a luminous moss is abundant; it is also found in the mines of Hesse. According to Heinzmann, the *rhizomorpha subterranea* and *aidulæ* are also phosphorescent.—See *Poetry of Science*.

PHOSPHORESCENCE OF THE SEA.

By microscopic examination of the myriads of minute insects which cause this phenomenon, no other fact has been elicited than that they contain a fluid which, when squeezed out, leaves a train of light upon the surface of the water. The creatures appear almost invariably on the eve of some change of weather, which would lead us to suppose that their luminous phenomena must be connected with electrical excitation; and of this Mr. C. Peach of Fowey has furnished the most satisfactory proofs yet obtained.13

LIGHT FROM THE JUICE OF A PLANT.

In Brazil has been observed a plant, conjectured to be an Euphorbium, very remarkable for the light which it yields when cut. It contains a milky juice, which exudes as soon as the plant is wounded, and appears luminous for several seconds.

LIGHT FROM FUNGUS.

Phosphorescent funguses have been found in Brazil by Mr. Gardner, growing on the decaying leaves of a dwarf palm. They vary from one to two inches across, and the whole plant gives out at night a bright phosphorescent light, of a pale greenish hue, similar to that emitted by fire-flies and phosphorescent marine animals. The light given out by a few of these fungi in a dark room is sufficient to read by. A very large phosphorescent species is occasionally found in the Swan River colony.

LIGHT FROM BUTTONS.

Upon highly polished gilt buttons no figure whatever can be seen by the most careful examination; yet, when they are made to reflect the light of the sun or of a candle upon a piece of paper held close to them, they give a beautiful geometrical figure, with ten rays issuing from the centre, and terminating in a luminous rim.

COLOURS OF SCRATCHES.

An extremely fine scratch on a well-polished surface may be regarded as having a concave, cylindrical, or at least a curved surface, capable of reflecting light in all directions; this is evident, for it is visible in all directions. Hence a single scratch or furrow in a surface may produce colours by the interference of the rays reflected from its opposite edges. Examine a spider's thread in the sunshine, and it will gleam with vivid colours. These may arise from a similar cause; or from the thread itself, as spun by the animal, consisting of several threads agglutinated together, and thus presenting, not a cylindrical, but a furrowed surface.

MAGIC BUST.

Sir David Brewster has shown how the rigid features of a white bust may be made to move and vary their expression, sometimes smiling and sometimes frowning, by moving rapidly in front of the bust a bright light, so as to make the lights and shadows take every possible direction and various degrees of intensity; and if the bust be placed before a concave mirror, its image may be made to do still more when it is cast upon wreaths of smoke.

COLOURS HIT MOST FREQUENTLY DURING BATTLE.

It would appear from numerous observations that soldiers are hit during battle according to the colour of their dress in the following order: red is the most fatal colour; the least fatal, Austrian gray. The proportions are, red, 12; rifle-green, 7; brown, 6; Austrian bluish-gray, 5.—*Jameson's Journal*, 1853.

TRANSMUTATION OF TOPAZ.

Yellow topazes may be converted into pink by heat; but it is a mistake to suppose that in the process the yellow colour is changed into pink: the fact is, that one of the pencils being yellow and the other pink, the yellow is discharged by heat, thus leaving the pink unimpaired.

COLOURS AND TINTS.

M. Chevreul, the *Directeur des Gobelins*, has presented to the French Academy a plan for a universal chromatic scale, and a methodical classification of all imaginable colours. Mayer, a professor at Göttingen, calculated that the different combinations of primitive colours produced 819 different tints; but M. Chevreul established not less than 14,424, all very distinct and easily recognised,—all of course proceeding from the three primitive simple colours of the solar spectrum, red, yellow, and blue. For example, he states that in the violet there are twenty-eight colours, and in the dahlia forty-two.

OBJECTS REALLY OF NO COLOUR.

A body appears to be of the colour which it reflects; as we see it only by reflected rays, it can but appear of the colour of those rays. Thus grass is green because it absorbs all except the green rays. Flowers, in the same manner, reflect the various colours of which they appear to us: the rose, the red rays; the violet, the blue; the daffodil, the yellow, &c. But these are not the permanent colours of the grass and flowers; for wherever you see these colours, the objects must be illuminated; and light, from whatever source it proceeds, is of the same nature, composed of the various coloured rays which paint the grass, the flowers, and every coloured object in nature. Objects in the dark have no colour, or are black, which is the same thing. You can never see objects without light. Light is composed of colours, therefore there can be no light without colours; and though every object is black or without colour in the dark, it becomes coloured as soon as it becomes visible.

THE DIORAMA—WHY SO PERFECT AN ILLUSION.

Because when an object is viewed at so great a distance that the optic axes of both eyes are sensibly parallel when directed towards it, the perspective projections of it, seen by each eye separately, are similar; and the appearance to the two eyes is precisely the same as when the object is seen by one eye only. There is, in such case, no difference between the visual appearance of an object in relief and its perspective projection on a plane surface; hence pictorial representations of distant objects, when those circumstances which would prevent or disturb the illusion are carefully excluded, may be rendered such perfect resemblances of the objects they are intended to represent as to be mistaken for them. The Diorama is an instance of this.— *Professor Wheatstone;Philosophical Transactions*, 1838.

CURIOUS OPTICAL EFFECTS AT THE CAPE.

Sir John Herschel, in his observatory at Feldhausen, at the base of the Table Mountain, witnessed several curious optical effects, arising from peculiar conditions of the atmosphere incident to the climate of the Cape. In the hot season "the nights are for the most part superb;" but occasionally, during the excessive heat and dryness of the sandy plains, "the optical tranquillity of the air" is greatly disturbed. In some cases, the images of the stars are violently dilated into nebular balls or puffs of 15′ in diameter; on other occasions they form "soft, round, quiet pellets of 3′ or 4′ diameter," resembling planetary nebulæ. In the cooler months the tranquillity of the image and the sharpness of vision are such, that hardly any limit is set to magnifying power but that which arises from the aberration of the specula. On occasions like these, optical phenomena of extraordinary splendour are produced by viewing a bright star through a diaphragm of cardboard or zinc pierced in regular patterns of circular holes by machinery: these phenomena surprise and delight every person that sees them. When close double stars are viewed with the telescope, with a diaphragm in the form of an equilateral triangle, the discs of the two stars, which are exact circles, have a clearness and perfection almost incredible.

THE TELESCOPE AND THE MICROSCOPE.

So singular is the position of the Telescope and the Microscope among the great inventions of the age, that no other process but that which they embody could make the slightest approximation to the secrets which they disclose. The steam-engine might have been imperfectly replaced by an air or an ether-engine; and a highly elastic fluid might have been, and may yet be, found, which shall impel the "rapid car," or drag the merchant-ship over the globe. The electric telegraph, now so perfect and unerring, might have spoken to us in the rude "language of chimes;" or sound, in place of electricity, might have passed along the metallic path, and appealed to the ear in place of the eye. For the printing-press and the typographic art might have been found a substitute, however poor, in the lithographic process; and knowledge might have been widely diffused by the photographic printing powers of the sun, or even artificial light. But without the telescope and the microscope, the human eye would have struggled in vain to study the worlds beyond our own, and the elaborate structures of the organic and inorganic creation could never have been revealed.— *North-British Review*, No. 50.

INVENTION OF THE MICROSCOPE.

The earliest magnifying lens of which we have any knowledge was one rudely made of rock-crystal, which Mr. Layard found, among a number of glass bowls, in the north-west palace of Nimroud; but no similar lens has been found or described to induce us to believe that the microscope, either single or compound, was invented and used as an instrument previous to the commencement of the seventeenth century. In the beginning of the first century, however, Seneca alludes to the magnifying power of a glass globe filled with water; but as he only states that it made small and indistinct letters appear larger and more distinct, we cannot consider such a casual remark as the invention of the single microscope, though it might have led the observer to try the effect of smaller globes, and thus obtain magnifying powers sufficient to discover phenomena otherwise invisible.

Lenses of glass were undoubtedly in existence at the time of Pliny; but at that period, and for many centuries afterwards, they appear to have been used only as burning or as reading glasses; and no attempt seems to have been made to form them of so small a size as to entitle them to be regarded even as the precursors of the single microscope.—*North-British Review*, No. 50.

The *rock-crystal lens* found at Nineveh was examined by Sir David Brewster. It was not entirely circular in its aperture. Its general form was that of a plano-convex lens, the plane side having been formed of one of the original faces of the six-sided crystal quartz, as Sir David ascertained by its action on polarised light: this was badly polished and scratched. The convex face of the lens had not been ground in a dish-shaped tool, in the manner in which lenses are now formed, but was shaped on a lapidary's wheel, or in some such manner. Hence it was unequally thick; but its extreme thickness was 2/10ths of an inch, its focal length being 4½ inches. It had twelve remains of cavities, which had originally contained liquids or condensed gases. Sir David has assigned reasons why this could not be looked upon as an ornament, but a true optical lens. In the same ruins were found some decomposed glass.

HOW TO MAKE THE FISH-EYE MICROSCOPE.

Very good microscopes may be made with the crystalline lenses of fish, birds, and quadrupeds. As the lens of fishes is spherical or spheroidal, it is absolutely necessary, previous to its use, to determine its optical axis and the axis of vision of the eye from which it is taken, and place the lens in such a manner that its axis is a continuation of the axis of our own eye. In no other direction but this is the albumen of which the lens consists symmetrically disposed in laminæ of equal density round a given line, which is the axis of the lens; and in no other direction does the gradation of density, by which the spherical aberration is corrected, preserve a proper relation to the axis of vision.

When the lens of any small fish, such as a minnow, a par, or trout, has been taken out, along with the adhering vitreous humour, from the eye-ball by cutting the sclerotic coat with a pair of scissors, it should be placed upon a piece of fine silver-paper previously freed from its minute adhering fibres. The absorbent nature of the paper will assist in removing all the vitreous humour from the lens; and when this is carefully done, by rolling it about with another piece of silver-paper, there will still remain, round or near the equator of the lens, a black ridge, consisting of the processes by which it was suspended in the eye-ball. The black circle points out to us the true axis of the lens, which is perpendicular to a plane passing through it. When the small crystalline has been freed from all the adhering vitreous humour, the capsule which contains it will have a surface as fine as a pellicle of fluid. It is then to be dropped from the paper into a cavity formed by a brass rim, and its position changed till the black circle is parallel to the circular rim, in which case only the axis of the lens will be a continuation of the axis of the observer's eye.—*Edin. Jour. Science*, vol. ii.

LEUWENHOECK'S MICROSCOPES.

Leuwenhoeck, the father of microscopical discovery, communicated to the Royal Society, in 1673, a description of the structure of a bee and a louse, seen by aid of his improved microscopes; and from this period until his decease in 1723, he regularly transmitted to the society his microscopical observations and discoveries, so that 375 of his papers and letters are preserved in the society's archives, extending over fifty years. He further bequeathed to the Royal Society a cabinet of twenty-six microscopes, which he had ground himself and set in silver, mostly extracted by him from minerals; these microscopes were exhibited to Peter the Great when he was at

Delft in 1698. In acknowledging the bequest, the council of the Royal Society, in 1724, presented Leuwenhoeck's daughter with a handsome silver bowl, bearing the arms of the society.—*Weld's History of the Royal Society*, vol. i.

DIAMOND LENSES FOR MICROSCOPES.

In recommending the employment of Diamond and other gems in the construction of Microscopes, Sir David Brewster has been met with the objection that they are too expensive for such a purpose; and, says Sir David, "they certainly are for instruments intended merely to instruct and amuse. But if we desire to make great discoveries, to unfold secrets yet hid in the cells of plants and animals, we must not grudge even a diamond to reveal them. If Mr. Cooper and Sir James South have given a couple of thousand pounds a piece for a refracting telescope, in order to study what have been miscalled 'dots' and 'lumps' of light on the sky; and if Lord Rosse has expended far greater sums on a reflecting telescope for analysing what has been called 'sparks of mud and vapour' encumbering the azure purity of the heavens,—why should not other philosophers open their purse, if they have one, and other noblemen sacrifice some of their household jewels, to resolve the microscopic structures of our own real world, and disclose secrets which the Almighty must have intended that we should know?"— *Proceedings of the British Association*, 1857.

THE EYE AND THE BRAIN SEEN THROUGH A MICROSCOPE.

By a microscopic examination of the retina and optic nerve and the brain, M. Bauer found them to consist of globules of 1/2800th to 1/4000th an inch diameter, united by a transparent viscid and coagulable gelatinous fluid.

MICROSCOPICAL EXAMINATION OF THE HAIR.

If a hair be drawn between the finger and thumb, from the end to the root, it will be distinctly felt to give a greater resistance and a different sensation to that which is experienced when drawn the opposite way: in consequence, if the hair be rubbed between the fingers, it will only move one way (travelling in the direction of a line drawn from its termination to its origin from the head or body), so that each extremity may thus be easily distinguished, even in the dark, by the touch alone.

The mystery is resolved by the achromatic microscope. A hair viewed on a dark ground as an *opaque* object with a high power, not less than that of a lens of one-thirtieth of an inch focus, and dully illuminated by a *cup*, the hair is seen to be indented with teeth somewhat resembling those of a coarse round rasp, but extremely irregular and rugged: as these incline all in one direction, like those of a common file, viz. from the origin of the hair towards its extremity, it sufficiently explains the above singular property.

This is a singular proof of the acuteness of the sense of feeling, for the said teeth may be felt much more easily than they can be seen. We may thus understand why a razor will cut a hair in two much more easily when drawn against its teeth than in the opposite direction.—*Dr. Goring.*

THE MICROSCOPE AND THE SEA.

What myriads has the microscope revealed to us of the rich luxuriance of animal life in the ocean, and conveyed to our astonished senses a consciousness of the universality of life! In the oceanic depths every stratum of water is animated, and swarms with countless hosts of small luminiferous animalcules, mammaria, crustacea, peridinea, and circling nereides, which, when attracted to the surface by peculiar meteorological conditions, convert every wave into a foaming band of flashing light.

USE OF THE MICROSCOPE TO MINERALOGISTS.

M. Dufour has shown that an imponderable quantity of a substance can be crystallised; and that the crystals so obtained are quite characteristic of the substances, as of sugar, chloride of sodium, arsenic, and mercury. This process may be extremely valuable to the mineralogist and toxicologist when the substance for examination is too small to be submitted to tests. By aid of the microscope, also, shells are measured to the thousandth part of an inch.

FINE DOWN OF QUARTZ.

Sir David Brewster having broken in two a crystal of quartz of a smoky colour, found both surfaces of the fracture absolutely black; and the blackness appeared at first sight to be owing to a thin film of opaque matter which had insinuated itself into the crevice. This opinion, however, was untenable, as every part of the surface was black, and the two halves of the crystals could not have stuck together had the crevice extended across the whole section. Upon further examination Sir David found that the surface was perfectly transparent by transmitted light, and that the blackness of the surfaces arose from their being entirely composed of *a fine down of quartz*, or of short and slender filaments, whose diameter was so exceedingly small that they were incapable of reflecting a single ray of the strongest light; and they could not exceed the *one third of the millionth part of an inch*. This curious specimen is in the cabinet of her grace the Duchess of Gordon.

MICROSCOPIC WRITING.

Professor Kelland has shown, in Paris, on a spot no larger than the head of a small pin, by means of powerful microscopes, several specimens of distinct and beautiful writing, one of them containing the whole of the Lord's Prayer written within this minute compass. In reference to this, two remarkable facts in Layard's latest work on Nineveh show that the national records of Assyria were written on square bricks, in characters so small as scarcely to be legible without a microscope; in fact, a microscope, as we have just shown, was found in the ruins of Nimroud.

HOW TO MAKE A MAGIC MIRROR.

Draw a figure with weak gum-water upon the surface of a convex mirror. The thin film of gum thus deposited on the outline or details of the figure will not be visible in dispersed daylight; but when made to reflect the rays of the sun, or those of a divergent pencil, will be beautifully displayed by the lines and tints occasioned by the diffraction of light, or the interference of the rays passing through the film with those which pass by it.

SIR DAVID BREWSTER'S KALEIDOSCOPE.

The idea of this instrument, constructed for the purpose of creating and exhibiting a variety of beautiful and perfectly symmetrical forms, first occurred to Sir David Brewster in 1814, when he was engaged in experiments on the polarisation of light by successive reflections between plates of glass. The reflectors were in some instances inclined to each other; and he had occasion to remark the circular arrangement of the images of a candle round a centre, or the multiplication of the sectors formed by the extremities of the glass plates. In repeating at a subsequent period the experiments of M. Biot on the action of fluids upon light, Sir David Brewster placed the fluids in a trough, formed by two plates of glass cemented together at an angle; and the eye being necessarily placed at one end, some of the cement, which had been pressed through between the plates, appeared to be arranged into a regular figure. The remarkable symmetry which it presented led to Dr. Brewster's investigation of the cause of this phenomenon; and in so doing he discovered the leading principles of the Kaleidoscope.

By the advice of his friends, Dr. Brewster took out a patent for his invention; in the specification of which he describes the kaleidoscope in two different forms. The instrument, however, having been shown to several opticians in London, became known before he could avail himself of his patent; and being simple in principle, it was at once largely manufactured. It is calculated that not less than 200,000 kaleidoscopes were sold in three months in London and Paris; though out of this number, Dr. Brewster says, not perhaps 1000 were constructed upon scientific principles, or were capable of giving any thing like a correct idea of the power of his kaleidoscope.

THE KALEIDOSCOPE THOUGHT TO BE ANTICIPATED.

In the seventh edition of a work on gardening and planting, published in 1739, by Richard Bradley, F.R.S., late Professor of Botany in the University of Cambridge, we find the following details of an invention, "by which the best designers and draughtsmen may improve and help their fancies. They must choose two pieces of looking-glass of equal bigness, of the figure of a long square. These must be covered on the back with paper or silk, to prevent rubbing off the silver. This covering must be so put on that nothing of it appears about the edges of the bright side. The glasses being thus prepared, must be laid face to face, and hinged together so that they may be made to open and shut at pleasure like the leaves of a book." After showing how various figures are to be looked at in these glasses under the same opening, and how the same figure may be varied under the different openings, the ingenious artist thus concludes: "If it should happen that the reader has any number of plans for parterres or wildernesses by him, he may by this method alter them at his pleasure, and produce such innumerable varieties as it is not possible the most able designer could ever have contrived."

MAGIC OF PHOTOGRAPHY.

Professor Moser of Königsberg has discovered that all bodies, even in the dark, throw out invisible rays; and that these bodies, when placed at a small distance from polished surfaces of all kinds, depict themselves upon such surfaces in forms which remain invisible till they are developed by the human breath or by the vapours of mercury or iodine. Even if the sun's image is made to pass over a plate of glass, the light tread of its rays will leave behind it an invisible track, which the human breath will instantly reveal.

Among the early attempts to take pictures by the rays of the sun was a very interesting and successful experiment made by Dr. Thomas Young. In 1802, when Mr. Wedgewood was "making profiles by the agency of light," and Sir Humphry Davy was "copying on prepared paper the images of small objects produced by means of the solar microscope," Dr. Young was taking photographs upon paper dipped in a solution of nitrate of silver, of the coloured rings observed by Newton; and his experiments clearly proved that the agent was not the luminous rays in the sun's light, but the invisible or chemical rays beyond the violet. This experiment is described in the Bakerian Lecture, 1803.

Niepce (says Mr. Hunt) pursued a physical investigation of the curious change, and found that all bodies were influenced by this principle radiated from the sun. Daguerre[14] produced effects from the solar pencil which no artist could approach; and Talbot and others extended the application. Herschel took up the inquiry; and he, with his usual power of inductive search and of philosophical deduction, presented the world with a class of discoveries which showed how vast a field of investigation was opening for the younger races of mankind.

The first attempts in photography, which were made at the instigation of M. Arago, by order of the French Government, to copy the Egyptian tombs and temples and the remains of the Aztecs in Central America, were failures. Although the photographers employed succeeded to admiration, in Paris, in producing pictures in a few minutes, they found often that an exposure of an hour was insufficient under the bright and glowing illumination of a southern sky.

THE BEST SKY FOR PHOTOGRAPHY.

Contrary to all preconceived ideas, experience proves that the brighter the sky that shines above the camera the more tardy the action within it. Italy and Malta do their work slower than Paris. Under the brilliant light of a Mexican sun, half an hour is required to produce effects which in England would occupy but a minute. In the burning atmosphere of India, though photographical the year round, the process is comparatively slow and difficult to manage; while in the clear, beautiful, and moreover cool, light of the higher Alps of Europe, it has been proved that the production of a picture requires many more minutes, even with the most sensitive preparations, than in the murky atmosphere of London. Upon the whole, the temperate skies of this country may be pronounced favourable to photographic action; a fact for which the prevailing characteristic of our climate may partially account, humidity being an indispensable condition for the working state both of paper and chemicals.— *Quarterly Review*, No. 202.

PHOTOGRAPHIC EFFECTS OF LIGHTNING.

The following authenticated instances of this singular phenomenon have been communicated to the Royal Society by Andrés Poey, Director of the Observatory at Havana:

Benjamin Franklin, in 1786, stated that about twenty years previous, a man who was standing opposite a tree that had just been struck by "a thunderbolt" had on his breast an exact representation of that tree.

In the New-York *Journal of Commerce*, August 26th, 1853, it is related that "a little girl was standing at a window, before which was a young maple-tree; after a brilliant flash of lightning, a complete image of the tree was found imprinted on her body."

M. Raspail relates that, in 1855, a boy having climbed a tree for the purpose of robbing a bird's nest, the tree was struck, and the boy thrown upon the ground; on his breast the image of the tree, with the bird and nest on one of its branches, appeared very plainly.

M. Olioli, a learned Italian, brought before the Scientific Congress at Naples the following four instances: 1. In September 1825, the foremast of a brigantine in the Bay of St. Arniro was struck by lightning, when a sailor sitting under the mast was struck dead, and on his back was found an impression of a horse-shoe, similar even in size to that fixed on the mast-head. 2. A sailor, standing in a similar position, was struck by lightning, and had on his left breast the impression of the number 4 4, with a dot between the two figures, just as they appeared at the extremity of one of the masts. 3. On the 9th October 1836, a young man was found struck by lightning; he had on a girdle, with some gold coins in it, which were imprinted on his skin in the order they were placed in the girdle,—a series of circles, with one point of contact, being plainly visible. 4. In 1847, Mme. Morosa, an Italian lady of Lugano, was sitting near a window during a thunderstorm, and perceived the commotion, but felt no injury; but a flower which happened to be in the path of the electric current was perfectly reproduced on one of her legs, and there remained permanently.

M. Poey himself witnessed the following instance in Cuba. On July 24th, 1852, a poplar-tree in a coffee-plantation was struck by lightning, and on one of the large dry leaves was found an exact representation of some pine-trees that lay 367 yards distant.

M. Poey considers these lightning impressions to have been produced in the same manner as the electric images obtained by Moser, Riess, Karster, Grove, Fox Talbot, and others, either by statical or dynamical electricity of different intensities. The fact that impressions are made through the garments is easily accounted for by their rough texture not preventing the lightning passing through them with the impression. To corroborate this view, M. Poey mentions an instance of lightning passing down a chimney into a trunk, in which was found an inch depth of soot, which must have passed through the wood itself.

PHOTOGRAPHIC SURVEYING.

During the summer of 1854, in the Baltic, the British steamers employed in examining the enemy's coasts and fortifications took photographic views for reference and minute examination. With the steamer moving at the rate of fifteen knots an hour, the most perfect definitions of coasts and batteries were obtained. Outlines of the coasts, correct in height and distance, have been faithfully transcribed; and all details of the fortresses passed under this photographic review are accurately recorded.

It is curious to reflect that the aids to photographic development all date within the last half-century, and are but little older than photography itself. It was not until 1811 that the chemical substance called iodine, on which the foundations of all popular photography rest, was discovered at all; bromine, the only other substance equally sensitive, not till 1826. The invention of the electro process was about simultaneous with that of photography itself. Gutta-percha only just preceded the substance of which collodion is made; the ether and chloroform, which are used in some methods, that of collodion. We say nothing of the optical improvements previously contrived or adapted for the purpose of the photograph: the achromatic lenses, which correct the discrepancy between the visual and chemical foci; the double lenses, which increase the force of the action; the binocular lenses, which do the work of the stereoscope; nor of the innumerable other mechanical aids which have sprung up for its use.

THE STEREOSCOPE AND THE PHOTOGRAPH.

When once the availability of one great primitive agent is worked out, it is easy to foresee how extensively it will assist in unravelling other secrets in natural science. The simple principle of the Stereoscope, for instance, might have been discovered a century ago, for the reasoning which led to it was independent of all the properties of light; but it could never have been illustrated, far less multiplied as it now is, without Photography. A few diagrams, of sufficient identity and difference to prove the truth of the principle, might have been constructed by hand, for the gratification of a few sages; but no artist, it is to be hoped, could have been found possessing the requisite ability and stupidity to execute the two portraits, or two groups, or two interiors, or two landscapes, identical in every minutia of the most elaborate detail, and yet differing in point of view by the inch between the two human eyes, by which the principle is brought to the level of any capacity. Here, therefore, the accuracy and insensibility of a machine could alone avail; and if in the order of things the cheap popular toy which the stereoscope now represents was necessary for the use of man, the photograph was first necessary for the service of the stereoscope.—*Quarterly Review*, No. 202.

THE STEREOSCOPE SIMPLIFIED.

When we look at any round object, first with one eye, and then with the other, we discover that with the right eye we see most of the right-hand side of the object, and with the left eye most of the left-hand side. These two images are combined, and we see an object which we know to be round.

This is illustrated by the *Stereoscope*, which consists of two mirrors placed each at an angle of 45 deg., or of two semi-lenses turned with their curved sides towards each other. To view its phenomena two pictures are obtained by the camera on photographic paper of any object in two positions, corresponding with the conditions of viewing it with the two eyes. By the mirrors on the lenses these dissimilar pictures are combined within the eye, and the vision of an actually solid object is produced from the pictures represented on a plane surface. Hence the name of the instrument, which signifies *Solid I see.—Hunt's Poetry of Science.*

PHOTO-GALVANIC ENGRAVING.

That which was the chief aid of Niepce in the humblest dawn of the art, viz. to transform the photographic plate into a surface capable of being printed, is in the above process done by the coöperation of Electricity with Photography. This invention of M. Pretsch, of Vienna, differs from all other attempts for the same purpose in not operating upon the photographic tablet itself, and by discarding the usual means of varnishes and bitings-in. The process is simply this: A glass tablet is coated with gelatine diluted till it forms a jelly, and containing bi-chromate of potash, nitrate of silver, and iodide of potassium. Upon this, when dry, is placed face downwards a paper positive, through which the light, being allowed to fall, leaves upon the gelatine a representation of the print. It is then soaked in water; and while the parts acted upon by the light are comparatively unaffected by the fluid, the remainder of the jelly swells, and rising above the general surface, gives a picture in relief, resembling an ordinary engraving upon wood. Of this intaglio a cast is now taken in gutta-percha, to which the electro process in copper being applied, a plate or matrix is produced, bearing on it an exact repetition of the original positive picture. All that now remains to be done is to repeat the electro process; and the result is a copper-plate in the necessary relievo, of which it has been said nature furnished the materials and science the artist, the inferior workman being only needed to roll it through the press.—*Quarterly Review*, No. 202.

SCIENCE OF THE SOAP-BUBBLE.

Few of the minor ingenuities of mankind have amused so many individuals as the blowing of bubbles with soap-lather from the bowl of a tobacco-pipe; yet how few who in childhood's careless hours have thus amused themselves, have in after-life become acquainted with the beautiful phenomena of light which the soap-bubble will enable us to illustrate!

Usually the bubble is formed within the bowl of a tobacco-pipe, and so inflated by blowing through the stem. It is also produced by introducing a capillary tube under the surface of soapy water, and so raising a bubble, which may be inflated to any convenient size. It is then guarded with a glass cover, to prevent its bursting by currents of air, evaporation, and other causes.

When the bubble is first blown, its form is elliptical, into which it is drawn by its gravity being resisted; but the instant it is detached from the pipe, and allowed to float in air, it becomes a perfect sphere, since the air within presses equally in all directions. There is also a strong cohesive attraction in the particles of soap and water, after having been forcibly distended; and as a sphere or globe possesses less surface than any other figure of equal capacity, it is of all forms the best adapted to the closest approximation of the particles of soap and water, which is another reason why the bubble is globular. The film of which the bubble consists is inconceivably thin (not exceeding the two-millionth part of an inch); and by the evaporation from its surface, the contraction and expansion of the air within, and the tendency of the soap-lather to gravitate towards the lower part of the bubble, and consequently to render the upper part still thinner, it follows that the bubble lasts but a few seconds. If, however, it were blown in a glass vessel, and the latter immediately closed, it might remain for some time; Dr. Paris thus preserved a bubble for a considerable period.

Dr. Hooke, by means of the coloured rings upon the soap-bubble, studied the curious subject of the colours of thin plates, and its application to explain the colours of natural bodies. Various phenomena were also discovered by Newton, who thus did not disdain to make a soap-bubble the object of his study. The colours which are reflected from the upper surface of the bubble are caused by the decomposition of the light which falls upon it; and the range of the phenomena is alike extensive and beautiful.15

Newton (says Sir D. Brewster), having covered the soap-bubble with a glass shade, saw its colours emerge in regular order, like so many concentric rings encompassing the top of it. As the bubble grew thinner by the continual subsidence of the water, the rings dilated slowly, and overspread the whole of it, descending to the bottom, where they vanished successively. When the colours had all emerged from the top, there arose in the centre of the rings a small round black spot, dilating it to more than half an inch in breadth till the bubble burst. Upon examining the rings between the object-glasses, Newton found that when they were only *eight* or *nine* in number, more than *forty* could be seen by viewing them through a prism; and even when the plate of air seemed all over uniformly white, multitudes of rings were disclosed by the prism. By means of these observations Newton was enabled to form his *Scale of Colours*, of great value in all optical researches.

Dr. Reade has thus produced a permanent soap-bubble:

Put into a six-ounce phial two ounces of distilled water, and set the phial in a vessel of water boiling on the fire. The water in the phial will soon boil, and steam will issue from its mouth, expelling the whole of the atmospheric air from within. Then throw in a piece of soap about the size of a small pea, cork the phial, and at the same instant remove it and the vessel from the fire. Then press the cork farther into the neck of the phial, and cover it thickly with sealing-wax; and when the contents are cold, a perfect vacuum will be formed within the bottle,—much better, indeed, than can be produced by the best-constructed air-pump.

To form a bubble, hold the bottle horizontally in both hands, and give it a sudden upward motion, which will throw the liquid into a wave, whose crest touching the upper interior surface of the phial, the tenacity of the liquid will cause a film to be retained all round the phial. Next place the phial on its bottom; when the film will form a section of the cylinder, being nearly but never quite horizontal. The film will be now colourless, since it reflects all the light which falls upon it. By remaining at rest for a minute or two, minute currents of lather will descend by their gravitating force down the inclined plane formed by the film, the upper part of which thus becomes drained to the necessary thinness; and this is the part to be observed.

Several concentric segments of coloured rings are produced; the colours, beginning from the top, being as follows:

1st order: Black, white, yellow, orange, red.

2d order: Purple, blue, white, yellow, red.

3d order: Purple, blue, green, yellowish-green, white, red.

4th order: Purple, blue, green, white, red.

5th order: Greenish-blue, very pale red.

6th order: Greenish-blue, pink.

7th order: Greenish-blue, pink.

As the segments advance they get broader, while the film becomes thinner and thinner. The several orders disappear upwards as the film becomes too thin to reflect their colours, until the first order alone remains, occupying the whole surface of the film. Of this order the red disappears first, then the orange, and lastly the yellow. The film is now divided by a line into two nearly equal portions, one black and the other white. This remains for some time; at length the film becomes too thin to hold together, and then vanishes. The colours are not faint and imperfect, but well defined, glowing with gorgeous hues, or melting into tints so exquisite as to have no rival through the whole circle of the arts. We quote these details from Mr. Tomlinson's excellent *Student's Manual of Natural Philosophy*.

We find the following anecdote related of Newton at the above period. When Sir Isaac changed his residence, and went to live in St. Martin's Street, Leicester Square, his next-door neighbour was a widow lady, who was much puzzled by the little she observed of the habits of the philosopher. A Fellow of the Royal Society called upon her one day, when, among her domestic news, she mentioned that some one had come to reside in the adjoining house who, she felt certain, was a poor crazy gentleman, "because," she continued, "he diverts himself in the oddest way imaginable. Every morning, when the sun shines so brightly that we are obliged to draw the window-blinds, he takes his seat on a little stool before a tub of soapsuds, and occupies himself for hours blowing soap-bubbles through a common clay-pipe, which bubbles he intently watches floating about till they burst. He is doubtless," she added, "now at his favourite amusement, for it is a fine day; do come and look at him." The gentleman smiled, and they went upstairs; when, after looking through the staircase-window into the adjoining court-yard, he turned and said: "My dear madam, the person whom you suppose to be a poor lunatic is no other than the great Sir Isaac Newton studying the refraction of light upon thin plates; a phenomenon which is beautifully exhibited on the surface of a common soap-bubble."

LIGHT FROM QUARTZ.

Among natural phenomena (says Sir David Brewster) illustrative of the colours of thin plates, we find none more remarkable than one exhibited by the fracture of a large crystal of quartz of a smoky colour, and about two and a quarter inches in diameter. The surface of fracture, in place of being a face or cleavage, or irregularly conchoidal, as we have sometimes seen it, was filamentous, like a surface of velvet, and consisted of short fibres, so small as to be incapable of reflecting light. Their size could not have been greater than the third of the millionth part of an inch, or one-fourth of the thinnest part of the soap-bubble when it exhibits the black spot where it bursts.

CAN THE CAT SEE IN THE DARK?

No, in all probability, says the reader; but the opposite popular belief is supported by eminent naturalists.

Buffon says: "The eyes of the cat shine in the dark somewhat like diamonds, which throw out during the night the light with which they were in a manner impregnated during the day."

Valmont de Bamare says: "The pupil of the cat is during the night still deeply imbued with the light of the day;" and again, "the eyes of the cat are during the night so imbued with light that they then appear very shining and luminous."

Spallanzani says: "The eyes of cats, polecats, and several other animals, shine in the dark like two small tapers;" and he adds that this light is phosphoric.

Treviranus says: "The eyes of the cat *shine where no rays of light penetrate*; and the light must in many, if not in all, cases proceed from the eye itself."

Now, that the eyes of the cat do shine in the dark is to a certain extent true: but we have to inquire whether by *dark* is meant the entire absence of light; and it will be found that the solution of this question will dispose of several assertions and theories which have for centuries perplexed the subject.

Dr. Karl Ludwig Esser has published in Karsten's Archives the results of an experimental inquiry on the luminous appearance of the eyes of the cat and other animals, carefully distinguishing such as evolve light from those which only reflect it. Having brought a cat into a half-darkened room, he observed from a certain direction that the cat's eyes, when *opposite the window*, sparkled brilliantly; but in other positions the light suddenly vanished. On causing the cat to be held so as to exhibit the light, and then gradually darkening the room, the light disappeared by the time the room was made quite dark.

In another experiment, a cat was placed opposite the window in a darkened room. A few rays were permitted to enter, and by adjusting the light, one or both of the cat's eyes were made to shine. In proportion as the pupil was dilated, the eyes were brilliant. By suddenly admitting a strong glare of light into the room, the pupil contracted; and then suddenly darkening the room, the eye exhibited a small round luminous point, which enlarged as the pupil dilated.

The eyes of the cat sparkle most when the animal is in a lurking position, or in a state of irritation. Indeed, the eyes of all animals, as well as of man, appear brighter when in rage than in a quiescent state, which Collins has commemorated in his Ode on the Passions:

"Next Anger rushed, his eyes on fire."

This brilliancy is said to arise from an increased secretion of the lachrymal fluid on the surface of the eye, by which the reflection of the light is increased. Dr. Esser, in places absolutely dark, never discovered the slightest trace of light in the eye of the cat; and he has no doubt that in all cases where cats' eyes have been seen to shine in dark places, such as a cellar, light penetrated through some window or aperture, and fell upon the eyes of the animal as it turned towards the opening, while the observer was favourably situated to obtain a view of the reflection.

To prove more clearly that this light does not depend upon the will of the animal, nor upon its angry passions, experiments were made upon the head of a dead cat. The sun's rays were admitted through a small aperture; and falling immediately upon the eyes, caused them to glow with a beautiful green light more vivid even than in the case of a living animal, on account of the increased dilatation of the pupil. It was also remarked that black and fox-coloured cats gave a brighter light than gray and white cats.

To ascertain the cause of this luminous appearance Dr. Esser dissected the eyes of cats, and exposed them to a small regulated amount of light after having removed different portions. The light was not diminished by the removal of the cornea, but only changed in colour. The light still continued after the iris was displaced; but on taking away the crystalline lens it greatly diminished both in intensity and colour. Dr. Esser then conjectured that the tapetum in the hinder part of the eye must form a spot which caused the reflection of the incident rays of light, and thus produce the shining; and this appeared more probable as the light of the eye now seemed to emanate from a single spot. Having taken away the vitreous humour, Dr. Esser observed that the entire want of the pigment in the hinder part of the choroid coat, where the optic nerve enters, formed a greenish, silver-coloured, changeable oblong spot, which was not symmetrical, but surrounded the optic nerve so that the greater part was above and only the smaller part below it; wherefore the greater part lay beyond the axis of vision. It is this spot, therefore, that produces the reflection of the incident rays of light, and beyond all doubt, according to its tint, contributes to the different colouring of the light.

It may be as well to explain that the interior of the eye is coated with a black pigment, which has the same effect as the black colour given to the inner surface of optical instruments: it absorbs any rays of light that may be reflected within the eye, and prevents them from being thrown again upon the retina so as to interfere with the distinctness of the images formed upon it. The retina is very transparent; and if the surface behind it, instead of being of a dark colour, were capable of reflecting light, the luminous rays which had already acted on the retina would be reflected back again through it, and not only dazzle from excess of light, but also confuse and render indistinct the images formed on the retina. Now in the case of the cat this black pigment, or a portion of it, is wanting; and those parts of the eye from which it is absent, having either a white or a metallic lustre, are called the tapetum. The smallest portion of light entering from it is reflected as by a concave mirror; and hence it is that the eyes of animals provided with this structure are luminous in a very faint light.

These experiments and observations show that the shining of the eyes of the cat does not arise from a phosphoric light, but only from a reflected light; that consequently it is not an effect of the will of the animal, or of violent passions; that their shining does not appear in absolute darkness; and that it cannot enable the animal to move securely in the dark.

It has been proved by experiment that there exists a set of rays of light of far higher refrangibility than those seen in the ordinary Newtonian spectrum. Mr. Hunt considers it probable that these highly refrangible rays, although under ordinary circumstances invisible to the human eye, may be adapted to produce the necessary degree of excitement upon which vision depends in the optic nerves of the night-roaming animals. The bat, the owl, and the cat may see in the gloom of the night by the aid of rays which are invisible to, or inactive on, the eyes of man or those animals which require the light of day for perfect vision.

Astronomy.

THE GREAT TRUTHS OF ASTRONOMY.

The difficulty of understanding these marvellous truths has been glanced at by an old divine (see *Things not generally Known*, p. 1); but the rarity of their full comprehension by those unskilled in mathematical science is more powerfully urged by Lord Brougham in these cogent terms:

Satisfying himself of the laws which regulate the mutual actions of the planetary bodies, the mathematician can convince himself of a truth yet more sublime than Newton's discovery of gravitation, though flowing from it; and must yield his assent to the marvellous position, that all the irregularities occasioned in the system of the universe by the mutual attraction of its members are periodical, and subject to an eternal law, which prevents them from ever exceeding a stated amount, and secures through all time the balanced structure of a universe composed of bodies whose mighty bulk and prodigious swiftness of motion mock the utmost efforts of the human imagination. All these truths are to the skilful mathematician as thoroughly known, and their evidence is as clear, as the simplest proposition of arithmetic to common understandings. But how few are those who thus know and comprehend them! Of all the millions that thoroughly believe these truths, certainly not a thousand individuals are capable of following even any considerable portion of the demonstrations upon which they rest; and probably not a hundred now living have ever gone through the whole steps of these demonstrations.—*Dissertations on Subjects of Science connected with Natural Theology*, vol. ii.

Sir David Brewster thus impressively illustrates the same subject:

Minds fitted and prepared for this species of inquiry are capable of appreciating the great variety of evidence by which the truths of the planetary system are established; but thousands of individuals, and many who are highly distinguished in other branches of knowledge, are incapable of understanding such researches, and view with a sceptical eye the great and irrefragable truths of astronomy.

That the sun is stationary in the centre of our system; that the earth moves round the sun, and round its own axis; that the diameter of the earth is 8000 miles, and that of the sun *one hundred and ten times as great*; that the earth's orbit is 190,000,000 of miles in breadth; and that if this immense space were filled with light, it would appear only like a luminous point at the nearest fixed star,—are positions absolutely unintelligible and incredible to all who have not carefully studied the subject. To millions of our species, then, the great Book of Nature is absolutely sealed; though it is in the power of all to unfold its pages, and to peruse those glowing passages which proclaim the power and wisdom of its Author.

ASTRONOMY AND DATES ON MONUMENTS.

Astronomy is a useful aid in discovering the Dates of ancient Monuments. Thus, on the ceiling of a portico among the ruins of Tentyris are the twelve signs of the Zodiac, placed according to the apparent motion of the sun. According to this Zodiac, the summer solstice is in Leo; from which it is easy to compute, by the precession of the equinoxes of 50″•1 annually, that the Zodiac of Tentyris must have been made 4000 years ago.

Mrs. Somerville relates that she once witnessed the ascertainment of the date of a Papyrus by means of astronomy. The manuscript was found in Egypt, in a mummy-case; and its antiquity was determined by the configuration of the heavens at the time of its construction. It proved to be a horoscope of the time of Ptolemy.

"THE CRYSTAL VAULT OF HEAVEN."

This poetic designation dates back as far as the early period of Anaximenes; but the first clearly defined signification of the idea on which the term is based occurs in Empedocles. This philosopher regarded the heaven of the fixed stars as a solid mass, formed from the ether which had been rendered crystalline by the action of fire.

In the Middle Ages, the fathers of the Church believed the firmament to consist of from seven to ten glassy strata, incasing each other like the different coatings of an onion. This supposition still keeps its ground in some of the monasteries of southern Europe, where Humboldt was greatly surprised to hear a venerable prelate express an opinion in reference to the fall of aerolites at Aigle, that the bodies we called meteoric stones with vitrified crusts were not portions of the fallen stone itself, but simply fragments of the crystal vault shattered by it in its fall.

Empedocles maintained that the fixed stars were riveted to the crystal heavens; but that the planets were free and unconstrained. It is difficult to conceive how, according to Plato in the *Timæus*, the fixed stars, riveted as they are to solid spheres, could rotate independently.

Among the ancient views, it may be mentioned that the equal distance at which the stars remained, while the whole vault of heaven seemed to move from east to west, had led to the idea of a firmament and a solid crystal sphere, in which Anaximenes (who was probably not much later than Pythagoras) had conjectured that the stars were riveted like nails.

MUSIC OF THE SPHERES.

The Pythagoreans, in applying their theory of numbers to the geometrical consideration of the five regular bodies, to the musical intervals of tone which determine a word and form different kinds of sounds, extended it even to the system of the universe itself; supposing that the moving, and, as it were, vibrating planets, exciting sound-waves, must produce a *spheral music*, according to the harmonic relations of their intervals of space. "This music," they add, "would be perceived by the human ear, if it was not rendered insensible by extreme familiarity, as it is perpetual, and men are accustomed to it from childhood."

The Pythagoreans affirm, in order to justify the reality of the tones produced by the revolution of the spheres, that hearing takes place only where there is an alternation of sound and silence. The inaudibility of the spheral music is also accounted for by its overpowering the senses. Aristotle himself calls the Pythagorean tone-myth pleasing and ingenious, but untrue.

Plato attempted to illustrate the tones of the universe in an agreeable picture, by attributing to each of the planetary spheres a syren, who, supported by the stern daughters of Necessity, the three Fates, maintain the eternal revolution of the world's axis. Mention is constantly made of the harmony of the spheres, though generally reproachfully, throughout the writings of Christian antiquity and the Middle Ages, from Basil the Great to Thomas Aquinas and Petrus Alliacus.

At the close of the sixteenth century, Kepler revived these musical ideas, and sought to trace out the analogies between the relations of tone and the distances of the planets; and Tycho Brahe was of opinion that the revolving conical bodies were capable of vibrating the celestial air (what we now call "resisting medium") so as to produce tones. Yet Kepler, although he had talked of Venus and the Earth sounding sharp in aphelion and flat in perihelion, and the highest tone of Jupiter and that of Venus coinciding in flat accord, positively declared there to be "no such things as sounds among the heavenly bodies, nor is their motion so turbulent as to elicit noise from the attrition of the celestial air." (See *Things not generally Known*, p. 44.)

"MORE WORLDS THAN ONE."

Although this opinion was maintained incidentally by various writers both on astronomy16 and natural religion, yet M. Fontenelle was the first individual who wrote a treatise on the *Plurality of Worlds*, which appeared in 1685, the year before the publication of Newton's *Principia*. Fontenelle's work consists of five chapters: 1. The earth is a planet which turns round its axis, and also round the sun. 2. The moon is a habitable world. 3. Particulars concerning the world in the moon, and that the other planets are also inhabited. 4. Particulars of the worlds of Venus, Mercury, Mars, Jupiter, and Saturn. 5. The fixed stars are as many suns, each of which illuminates a world. In a future edition, 1719, Fontenelle added, 6. New thoughts which confirm those in the preceding conversations, and the latest discoveries which have been made in the heavens. The next work on the subject was the *Theory of the Universe, or Conjectures concerning the Celestial Bodies and their Inhabitants*, 1698, by Christian Huygens, the contemporary of Newton.

The doctrine is maintained by almost all the distinguished astronomers and writers who have flourished since the true figure of the earth was determined. Giordano Bruna of Nola, Kepler, and Tycho Brahe, believed in it; and Cardinal Cusa and Bruno, before the discovery of binary systems among the stars, believed also that the stars were inhabited. Sir Isaac Newton likewise adopted the belief; and Dr. Bentley, Master of Trinity College, Cambridge, in his eighth sermon on the Confutation of Atheism from the origin and frame of the world, has ably maintained the same doctrine. In our own day we may number among its supporters the distinguished names of the Marquis de la Place, Sir William and Sir John Herschel, Dr. Chalmers, Isaac Taylor, and M. Arago. Dr. Chalmers maintains the doctrine in his *Astronomical Discourses*, which one Alexander Maxwell (who did not believe in the grand truths of astronomy) attempted to controvert, in 1820, in a chapter of a volume entitled *Plurality of Worlds*.

Next appeared *Of a Plurality of Worlds*, attributed to the Rev. Dr. Whewell, Master of Trinity College, Cambridge; urging the theological not less than the scientific reasons for believing in the old tradition of a single world, and maintaining that "the earth is really the largest planetary body in the solar system,—its domestic hearth, and the only world in the universe." "I do not pretend," says Dr. Whewell, "to disprove the plurality of worlds; but I ask in vain for any argument which makes the doctrine probable." "It is too remote from knowledge to be either proved or disproved." Sir David Brewster has replied to Dr. Whewell's Essay, in *More Worlds than One, the Creed of the Philosopher and the Hope of the Christian*, emphatically maintaining that analogy strongly countenances the idea of all the solar planets, if not all worlds in the universe, being peopled with creatures not dissimilar in being and nature to the inhabitants of the earth. This view is supported in*Scientific Certainties of Planetary Life*, by T. C. Simon, who well treats one point of the argument—that mere distance of the planets from the central sun does not determine the condition as to light and heat, but that the density of the ethereal medium enters largely into the calculation. Mr. Simon's general conclusion is, that "neither on account of deficient or excessive heat, nor with regard to the density of the materials, nor with regard to the force of gravity on the surface, is there the slightest pretext for supposing that all the planets of our system are not inhabited by intellectual creatures with animal bodies like ourselves,—moral beings, who know and love their great Maker, and who wait, like the rest of His creation, upon His providence and upon His care." One of the leading points of Dr. Whewell's Essay is, that we should not elevate the conjectures of analogy into the rank of scientific certainties; and that "the force of all the presumptions drawn from physical reasoning for the opinion of planets and stars being either inhabited or uninhabited is so small, that the belief of all thoughtful persons on this subject will be determined by moral, metaphysical, and theological considerations."

WORLDS TO COME—ABODES OF THE BLEST.

Sir David Brewster, in his eloquent advocacy of the doctrine of "more worlds than one," thus argues for their peopling:

Man, in his future state of existence, is to consist, as at present, of a spiritual nature residing in a corporeal frame. He must live, therefore, upon a material planet, subject to all the laws of matter, and performing functions for which a material body is indispensable. We must consequently find for the race of Adam, if not races that may have preceded him, a material home upon which they may reside, or by which they may travel, by means unknown to us, to other localities in the universe. At the present hour, the inhabitants of the earth are nearly *a thousand millions*; and by whatever process we may compute the numbers that have existed before the present generation, and estimate those that are yet to inherit the earth, we shall obtain a population which the habitable parts of our globe could not possibly accommodate. If there is not room, then, on our earth for the millions of millions of beings who have lived and died upon its surface, and who may yet live and die during the period fixed for its occupation by man, we can scarcely doubt that their future abode must be on some of the primary or secondary planets of the solar system, whose inhabitants have ceased to exist like those on the earth, or upon planets in our own or in other systems which have been in a state of preparation, as our earth was, for the advent of intellectual life.

"GAUGING THE HEAVENS."

Sir William Herschel, in 1785, conceived the happy idea of counting the number of stars which passed at different heights and in various directions over the field of view, of fifteen minutes in diameter, of his twenty-feet reflecting telescope. The field of view each time embraced only 1/833000th of the whole heavens; and it would therefore require, according to Struve, eighty-three years to gauge the whole sphere by a similar process.

VELOCITY OF THE SOLAR SYSTEM.

M. F. W. G. Struve gives as the splendid result of the united studies of MM. Argelander, O. Struve, and Peters, grounded on observations made at the three Russian observatories of Dorpat, Abo, and Pulkowa, "that the velocity of the motion of the solar system in space is such that the sun, with all the bodies which depend upon it, advances annually towards the constellation Hercules171•623 times the radius of the earth's orbit, or 33,550,000 geographical miles. The possible error of this last number amounts to 1,733,000 geographical miles, or to a *seventh* of the whole value. We may, then, wager 400,000 to 1 that the sun has a proper progressive motion, and 1 to 1 that it is comprised between the limits of thirty-eight and twenty-nine millions of geographical miles."

That is, taking 95,000,000 of English miles as the mean radius of the Earth's orbit, we have 95 × 1•623 = 154•185 millions of miles; and consequently,

English Miles.

The velocity of the Solar System 154,185,000 in the year.

" "   422,424 in a day.

" "   17,601 in an hour.

" "   293 in a minute.

" "   57 in a second.

The Sun and all his planets, primary and secondary, are therefore now in rapid motion round an invisible focus. To that now dark and mysterious centre, from which no ray, however feeble, shines, we may in another age point our telescopes, detecting perchance the great luminary which controls our system and bounds its path: into that vast orbit man, during the whole cycle of his race, may never be allowed to round.—*North-British Review*, No. 16.

NATURE OF THE SUN.

M. Arago has found, by experiments with the polariscope, that the light of gaseous bodies is natural light when it issues from the burning surface; although this circumstance does not prevent its subsequent complete polarisation, if subjected to suitable reflections or refractions. Hence we obtain *a most simple method of discovering the nature of the sun* at a distance of forty millions of leagues. For if the light emanating from the margin of the sun, and radiating from the solar substance *at an acute angle*, reach us without having experienced any sensible reflections or refractions in its passage to the earth, and if it offer traces of polarisation, the sun must be *a solid or a liquid body*. But if, on the contrary, the light emanating from the sun's margin give no indications of polarisation, the *incandescent* portion of the sun must be *gaseous*. It is by means of such a methodical sequence of observations that we may acquire exact ideas regarding the physical constitution of the sun.—*Note to Humboldt's Cosmos*, vol. iii.

STRUCTURE OF THE LUMINOUS DISC OF THE SUN.

The extraordinary structure of the *fully luminous* Disc of the Sun, as seen through Sir James South's great achromatic, in a drawing made by Mr. Gwilt, resembles compressed curd, or white almond-soap, or a mass of asbestos fibres, lying in a *quaquaversus* direction, and compressed into a solid mass. There can be no illusion in this phenomenon; it is seen by every person with good vision, and on every part of the sun's luminous surface or envelope, which is thus shown to be not a *flame*, but a soft solid or thick fluid, maintained in an incandescent state by subjacent heat, capable of being disturbed by differences of temperature, and broken up as we see it when the sun is covered with spots or openings in the luminous matter.—*North-British Review*, No. 16.

Copernicus named the sun the lantern of the world (*lucerna mundi*); and Theon of Smyrna called it the heart of the universe. The mass of the sun is, according to Encke's calculation of Sabine's pendulum formula, 359,551 times that of the earth, or 355,499 times that of the earth and moon together; whence the density of the sun is only about ¼ (or more accurately 0•252) that of the earth. The volume of the sun is 600 times greater, and its mass, according to Galle, 738 times greater, than that of all the planets combined. It may assist the mind in conceiving a sensuous image of the magnitude of the sun, if we remember that if the solar sphere were entirely hollowed out, and the earth placed in its centre, there would still be room enough for the moon to describe its orbit, even if the radius of the latter were increased 160,000 geographical miles. A railway-engine, moving at the rate of thirty miles an hour, would require 360 years to travel from the earth to the sun. The diameter of the sun is rather more than one hundred and eleven times the diameter of the earth. Therefore the volume or bulk of the sun must be nearly *one million four hundred thousand* times that of the earth. Lastly, if all the bodies composing the solar system were formed into one globe, it would be only about the five-hundredth part of the size of the sun.

GREAT SIZE OF THE SUN ON THE HORIZON EXPLAINED.

The dilated size (generally) of the Sun or Moon, when seen near the horizon, beyond what they appear to have when high up in the sky, has nothing to do with refraction. It is an illusion of the judgment, arising from the terrestrial objects interposed, or placed in close comparison with them. In that situation we view and judge of them as we do of terrestrial objects—in detail, and with an acquired attention to parts. Aloft we have no association to guide us, and their insulation in the expanse of the sky leads us rather to undervalue than to over-rate their apparent magnitudes. Actual measurement with a proper instrument corrects our error, without, however, dispelling our illusion. By this we learn that the sun, when just on the horizon, subtends at our eyes almost exactly the same, and the moon a materially *less*, angle than when seen at a greater altitude in the sky, owing to its greater distance from us in the former situation as compared with the latter.—*Sir John Herschel's Outlines.*

TRANSLATORY MOTION OF THE SUN.

This phenomenon is the progressive motion of the centre of gravity of the whole solar system in universal space. Its velocity, according to Bessel, is probably four millions of miles daily, in a *relative* velocity to that of 61 Cygni of at least 3,336,000 miles, or more than double the velocity of the revolution of the earth in her orbit round the sun. This change of the entire solar system would remain unknown to us, if the admirable exactness of our astronomical instruments of measurement, and the advancement recently made in the art of observing, did not cause our progress towards remote stars to be perceptible, like an approximation to the objects of a distant shore in apparent motion. The proper motion of the star 61 Cygni, for instance, is so considerable, that it has amounted to a whole degree in the course of 700 years.—*Humboldt's Cosmos*, vol. i.

THE SUN'S LIGHT COMPARED WITH TERRESTRIAL LIGHTS.

Mr. Ponton has by means of a simple monochromatic photometer ascertained that a small surface, illuminated by mean solar light, is 444 times brighter than when it is illuminated by a moderator lamp, and 1560 times brighter than when it is illuminated by a wax-candle (short six in the lb.)—the artificial light being in both instances placed at two inches' distance from the illuminated surface. And three electric lights, each equal to 520 wax-candles, will render a small surface as bright as when it is illuminated by mean sunshine.

It is thence inferred, that a stratum occupying the entire surface of the sphere of which the earth's distance from the sun is the radius, and consisting of three layers of flame, each 1/1000th of an inch in thickness, each possessing a brightness equal to that of such an electric light, and all three embraced within a thickness of 1/40th of an inch, would give an amount of illumination equal in quantity and intensity to that of the sun at the distance of 95 millions of miles from his centre.

And were such a stratum transferred to the surface of the sun, where it would occupy 46,275 times less area, its thickness would be increased to 94 feet, and it would embrace 138,825 layers of flame, equal in brightness to the electric light; but the same effect might be produced by a stratum about nine miles in thickness, embracing 72 millions of layers, each having only a brightness equal to that of a wax-candle.18

ACTINIC POWER OF THE SUN.

Mr. J. J. Waterston, in 1857, made at Bombay some experiments on the photographic power of the sun's direct light, to obtain data in an inquiry as to the possibility of measuring the diameter of the sun to a very minute fraction of a second, by combining photography with the principle of the electric telegraph; the first to measure the element space, the latter the element time. The result is that about 1/20000th of a second is sufficient exposure to the direct light of the sun to obtain a distinct mark on a sensitive collodion plate, when developed by the usual processes; and the duration of the sun's full action on any one point is about 1/9000th of a second.

M. Schatt, a young painter of Berlin, after 1500 experiments, succeeded in establishing a scale of all the shades of black which the action of the sun produces on photographic paper; so that by comparing the shade obtained at any given moment on a certain paper with that indicated on the scale, the exact force of the sun's light may be determined.

HEATING POWER OF THE SUN.

All moving power has its origin in the rays of the sun. While Stephenson's iron tubular railway-bridge over the Menai Straits, 400 feet long, bends but half an inch under the heaviest pressure of a train, it will bend up an inch and a half from its usual horizontal line when the sun shines on it for some hours. The Bunker-Hill monument, near Boston, U.S., is higher in the evening than in the morning of a sunny day; the little sunbeams enter the pores of the stone like so many wedges, lifting it up.

In winter, the Earth is nearer the Sun by about 1/30 than in summer; but the rays strike the northern hemisphere more obliquely in winter than the other half year.

M. Pouillet has estimated, with singular ingenuity, from a series of observations made by himself, that the whole quantity of heat which the Earth receives annually from the

Sun is such as would be sufficient to melt a stratum of ice covering the entire globe forty-six feet deep.

By the action of the sun's rays upon the earth, vegetables, animals, and man, are in their turn supported; the rays become likewise, as it were, a store of heat, and "the sources of those great deposits of dynamical efficiency which are laid up for human use in our coal strata" (*Herschel*).

A remarkable instance of the power of the sun's rays is recorded at Stonehouse Point, Devon, in the year 1828. To lay the foundation of a sea-wall the workmen had to descend in a diving-bell, which was fitted with convex glasses in the upper part, by which, on several occasions in clear weather, the sun's rays were so concentrated as to burn the labourers' clothes when opposed to the focal point, and this when the bell was twenty-five feet under the surface of the water!

CAUSE OF DARK COLOUR OF THE SKIN.

Darkness of complexion has been attributed to the sun's power from the age of Solomon to this day,—"Look not upon me, because I am black, because the sun hath looked upon me:" and there cannot be a doubt that, to a certain degree, the opinion is well founded. The invisible rays in the solar beams, which change vegetable colour, and have been employed with such remarkable effect in the daguerreotype, act upon every substance on which they fall, producing mysterious and wonderful changes in their molecular state, man not excepted.—*Mrs. Somerville.*

EXTREME SOLAR HEAT.

The fluctuation in the sun's direct heating power amounts to 1/15th, which is too considerable a fraction of the whole intensity not to aggravate in a serious degree the sufferings of those who are exposed to it in thirsty deserts without shelter. The amount of these sufferings, in the interior of Australia for instance, are of the most frightful kind, and would seem far to exceed what have ever been undergone by travellers in the northern deserts of Africa. Thus Captain Sturt, in his account of his Australian exploration, says: "The ground was almost a molten surface; and if a match accidentally fell upon it, it immediately ignited." Sir John Herschel has observed the temperature of the surface soil in South Africa as high as 159° Fahrenheit. An ordinary lucifer-match does not ignite when simply pressed upon a smooth surface at 212°; but *in the act of withdrawing it* it takes fire, and the slightest friction upon such a surface of course ignites it.

HOW DR. WOLLASTON COMPARED THE LIGHT OF THE SUN AND THE FIXED STARS.

In order to compare the Light of the Sun with that of a Star, Dr. Wollaston took as an intermediate object of comparison the light of a candle reflected from a bulb about a quarter of an inch in diameter, filled with quicksilver; and seen by one eye through a lens of two inches focus, at the same time that the star on the sun's image, *placed at a proper distance*, was viewed by the other eye through a telescope. The mean of various trials seemed to show that the light of Sirius is equal to that of the sun seen in a glass bulb 1/10th of an inch in diameter, at the distance of 210 feet; or that they are in the proportion of one to ten thousand millions: but as nearly one half of this light is lost by reflection, the real proportion between the light from Sirius and the sun is not greater than that of one to twenty thousand millions.

"THE SUN DARKENED."

Humboldt selects the following example from historical records as to the occurrence of a sudden decrease in the light of the Sun:

A.D. 33, the year of the Crucifixion. "Now from the sixth hour there was darkness over all the land till the ninth hour" (*St. Matthew* xxvii. 45). According to *St. Luke* (xxiii. 45), "the sun was darkened." In order to explain and corroborate these narrations, Eusebius brings forward an eclipse of the sun in the 202d Olympiad, which had been noticed by the chronicler Phlegon of Tralles (*Ideler, Handbuch der Mathem. Chronologie*, Bd. ii. p. 417). Wurn, however, has shown that the eclipse which occurred during this Olympiad, and was visible over the whole of Asia Minor, must have happened as early as the 24th of November 29 A.D. The day of the Crucifixion corresponded with the Jewish Passover (*Ideler*, Bd. i. pp. 515–520), on the 14th of the month Nisan, and the Passover was always celebrated at the time of the *full moon*. The sun cannot therefore have been darkened for three hours by the moon. The Jesuit Scheiner thinks the decrease in the light might be ascribed to the occurrence of large sun-spots.

THE SUN AND TERRESTRIAL MAGNETISM.

The important influence exerted by the Sun's body, as a mass, upon Terrestrial Magnetism, is confirmed by Sabine in the ingenious observation, that the period at which the intensity of the magnetic force is greatest, and the direction of the needle most near to the vertical line, falls in both hemispheres between the months of October and February; that is to say, precisely at the time when the earth is nearest to the sun, and moves in its orbit with the greatest velocity.

IS THE HEAT OF THE SUN DECREASING?

The Heat of the Sun is dissipated and lost by radiation, and must be progressively diminished unless its thermal energy be supplied. According to the measurements of M. Pouillet, the quantity of heat given out by the sun in a year is equal to that which would be produced by the combustion of a stratum of coal seventeen miles in thickness; and if the sun's capacity for heat be assumed equal to that of water, and the heat be supposed drawn uniformly from its entire mass, its temperature would thereby undergo a diminution of 20•4° Fahr. annually. On the other hand, there is a vast store of force in our system capable of conversion into heat. If, as is indicated by the small density of the sun, and by other circumstances, that body has not yet reached the condition of incompressibility, we have in the future approximation of its parts a fund of heat, probably quite large enough to supply the wants of the human family to the end of its sojourn here. It has been calculated that an amount of condensation which would diminish the diameter of the sun by only the ten-thousandth part, would suffice to restore the heat emitted in 2000 years.

UNIVERSAL SUN-DIAL.

Mr. Sharp, of Dublin, exhibited to the British Association in 1849 a Dial, consisting of a cylinder set to the day of the month, and then elevated to the latitude. A thin plane of metal, in the direction of its axis, is then turned by a milled head below it till the shadow is a minimum, when a dial on the top shows the hours by one hand, and the minutes by another, to the precision of about three minutes.

LENGTH OF DAYS AT THE POLES.

During the summer, in the northern hemisphere, places near the North Pole are in *continual sunlight*—the sun never sets to them; while during that time places near the South Pole never see the sun. When it is summer in the southern hemisphere, and the sun shines on the South Pole without setting, the North Pole is entirely deprived of his light. Indeed, at the Poles there is but *one day and one night*; for the sun shines for six months together on one Pole, and the other six months on the other Pole.

HOW THE DISTANCE OF THE SUN IS ASCERTAINED BY THE YARD-MEASURE.

Professor Airy, in his *Six Lectures on Astronomy*, gives a masterly analysis of a problem of considerable intricacy, viz. the determination of the parallax of the sun, and consequently of his distance, by observations of the transit of Venus, the connecting link between measures upon the earth's surface and the dimensions of our system. The further step of investigating the parallax, and consequently the distance of the fixed stars (where that is practicable), is also elucidated; and the author, with evident satisfaction, thus sums up the several steps:

By means of a yard-measure, a base-line in a survey was measured; from this, by the triangulations and computations of a survey, an arc of meridian on the earth was measured; from this, with proper observations with the zenith sector, the surveys being also repeated on different parts of the earth, the earth's form and dimensions were ascertained; from these, and a previous independent knowledge of the proportions of the distances of the earth and other planets from the sun, with observations of the transit of Venus, the sun's distance is determined; and from this, with observations leading to the parallax of the stars, the distance of the stars is determined. And *every step in the process can be distinctly referred to its basis, that is, the yard-measure.*

HOW THE TIDES ARE PRODUCED BY THE SUN AND MOON.

Each of these bodies excites, by its attraction upon the waters of the sea, two gigantic waves, which flow in the same direction round the world as the attracting bodies themselves apparently do. The two waves of the moon, on account of her greater nearness, are about 3½ times as large as those excited by the sun. One of these waves has its crest on the quarter of the earth's surface which is turned towards the moon; the other is at the opposite side. Both these quarters possess the flow of the tide, while the regions which lie between have the ebb. Although in the open sea the height of the tide amounts to only about three feet, and only in certain narrow channels, where the moving water is squeezed together, rises to thirty feet, the might of the phenomenon is nevertheless manifest from the calculation of Bessel, according to which a quarter of the earth covered by the sea possesses during the flow of the tide about 25,000 cubic miles of water more than during the ebb; and that, therefore, such a mass of water must in 6¼ hours flow from one quarter of the earth to the other.—*Professor Helmholtz.*

SPOTS ON THE SUN.

Sir John Herschel describes these phenomena, when watched from day to day, or even from hour to hour, as appearing to enlarge or contract, to change their forms, and at length disappear altogether, or to break out anew in parts of the surface where none were before. Occasionally they break up or divide into two or more. The scale on which their movements takes place is immense. A single second of angular measure, as seen from the earth, corresponds on the sun's disc to 461 miles; and a circle of this diameter (containing therefore nearly 167,000 square miles) is the least space which can be distinctly discerned on the sun as a *visible area.* Spots have been observed, however, whose linear diameter has been upwards of 45,000 miles; and even, if some records are to be trusted, of very much greater extent. That such a spot should close up in six weeks time (for they seldom last much longer), its borders must approach at the rate of more than 1000 miles a-day.

The same astronomer saw at the Cape of Good Hope, on the 29th March 1837, a solar spot occupying an area of near *five square minutes*, equal to 3,780,000,000 square miles. "The black centre of the spot of May 25th, 1837 (not the tenth part of the preceding one), would have allowed the globe of our earth to drop through it, leaving a thousand miles clear of contact on all sides of that tremendous gulf." For such an amount of disturbance on the sun's atmosphere, what reason can be assigned?

The Rev. Mr. Dawes has invented a peculiar contrivance, by means of which he has been enabled to scrutinise, under high magnifying power, minute portions of the solar disc. He places a metallic screen, pierced with a very small hole, in the focus of the telescope, where the image of the sun is formed. A small portion only of the image is thus allowed to pass through, so that it may be examined by the eye-piece without inconveniencing the observer by heat or glare. By this arrangement, Mr. Dawes has observed peculiarities in the constitution of the sun's surface which are discernible in no other way.

Before these observations, the dark spots seen on the sun's surface were supposed to be portions of the solid body of the sun, laid bare to our view by those immense fluctuations in the luminous regions of its atmosphere to which it appears to be subject. It now appears that these dark portions are only an additional and inferior stratum of a very feebly luminous or illuminated portion of the sun's atmosphere. This again in its turn Mr. Dawes has frequently seen pierced with a smaller and usually much more rounded aperture, which would seem at last to afford a view of the real solar surface of most intense blackness.

M. Schwabe, of Dessau, has discovered that the abundance or paucity of spots displayed by the sun's surface is subject to a law of periodicity. This has been confirmed by M. Wolf, of Berne, who shows that the period of these changes, from minimum to minimum, is 11 years and 11-hundredths of a year, being exactly at the rate of nine periods per century, the last year of each century being a year of minimum. It is strongly corroborative of the correctness both of M. Wolf's period and also of the periodicity itself, that of all the instances of the appearance of spots on the sun recorded in history, even before the invention of the telescope, or of remarkable deficiencies in the sun's light, of which there are great numbers, only two are found to deviate as much as two years from M. Wolf's epochs. Sir William Herschel observed that the presence or absence of spots had an influence on the temperature of the seasons; his observations have been fully confirmed by M. Wolf. And, from an examination of the chronicles of Zurich from A.D. 1000 to A.D. 1800, he has come to the conclusion "that years rich in solar spots are in general drier and more fruitful than those of an opposite character; while the latter are wetter and more stormy than the former."

The most extraordinary fact, however, in connection with the spots on the sun's surface, is the singular coincidence of their periods with those great disturbances in the magnetic system of the earth to which the epithet of "magnetic storms" has been affixed.

These disturbances, during which the magnetic needle is greatly and universally agitated (not in a particular limited locality, but at one and the same instant of time over whole continents, or even over the whole earth), are found, so far as observation has hitherto extended, to maintain a parallel, both in respect of their frequency of occurrence and intensity in successive years, with the abundance and magnitude of the spots in the same years, too close to be regarded as fortuitous. The coincidence of the epochs of maxima and minima in the two series of phenomena amounts, indeed, to identity; a fact evidently of most important significance, but which neither astronomical nor magnetic science is yet sufficiently advanced to interpret.—*Herschel's Outlines.*

The signification and connection of the above varying phenomena (Humboldt maintains) can never be manifested in their entire importance until an uninterrupted series of representations of the sun's spots can be obtained by the aid of mechanical clock-work and photographic apparatus, as the result of prolonged observations during the many months of serene weather enjoyed in a tropical climate.

M. Schwabe has thus distinguished himself as an indefatigable observer of the sun's spots, for his researches received the Royal Astronomical Society's Medal in 1857. "For thirty years," said the President at the presentation, "never has the sun exhibited his disc above the horizon of Dessau without being confronted by Schwabe's imperturbable telescope; and that appears to have happened on an average about 300 days a-year. So, supposing that he had observed but once a-day, he has made 9000 observations, in the course of which he discovered about 4700 groups. This is, I believe, an instance of devoted persistence unsurpassed in the annals of astronomy. The energy of one man has revealed a phenomenon that had eluded the suspicion of astronomers for 200 years."

HAS THE MOON AN ATMOSPHERE?

The Moon possesses neither Sea nor Atmosphere of appreciable extent. Still, as a negative, in such case, is relative only to the capabilities of the instruments employed, the search for the indications of a lunar atmosphere has been renewed with fresh augmentation of telescopic power. Of such indications, the most delicate, perhaps, are those afforded by the occultation of a planet by the moon. The occultation of Jupiter, which took place on January 2, 1857, was observed with this reference, and is said to have exhibited no *hesitation*, or change of form or brightness, such as would be produced by the refraction or absorption of an atmosphere. As respects the sea, if water existed on the moon's surface, the sun's light reflected from it should be completely polarised at a certain elongation of the moon from the sun; and no traces of such light have been observed.

MM. Baer and Maedler conclude that the moon is not entirely without an atmosphere, but, owing to the smallness of her mass, she is incapacitated from holding an extensive covering of gas; and they add, "it is possible that this weak envelope may sometimes, through local causes, in some measure dim or condense itself." But if any atmosphere exists on our satellite, it must be, as Laplace says, more attenuated than what is termed a vacuum in an air-pump.

Mr. Hopkins thinks that if there be any lunar atmosphere, it must be very rare in comparison with the terrestrial atmosphere, and inappreciable to the kind of observation by which it has been tested; yet the absence of any refraction of the light of the stars during occultation is a very refined test. Mr. Nasmyth observes that "the sudden disappearance of the stars behind the moon, without any change or diminution of her brilliancy, is one of the most beautiful phenomena that can be witnessed."

Sir John Herschel observes: The fact of the moon turning always the same face towards the earth is, in all probability, the result of an elongation of its figure in the direction of a line joining the centres of both the bodies, acting conjointly with a non-coincidence of its centre of gravity with its centre of symmetry.

If to this we add the supposition that the substance of the moon is not homogeneous, and that some considerable preponderance of weight is placed excentrically in it, it will be easily apprehended that the portion of its surface nearer to that heavier portion of its solid content, under all the circumstances of the moon's rotation, will permanently occupy the situation most remote from the earth.

In what regards its assumption of a definite level, air obeys precisely the same hydrostatical laws as water. The lunar atmosphere would rest upon the lunar ocean, and form in its basin a lake of air, whose upper portions at an altitude such as we are now contemplating would be of excessive tenuity, especially should the provision of air be less abundant in proportion than our own. It by no means follows, then, from the absence of visible indications of water or air on this side of the moon, that the other is equally destitute of them, and equally unfitted for maintaining animal or vegetable life. Some slight approach to such a state of things actually obtains on the earth itself. Nearly all the land is collected in one of its hemispheres, and much the larger portion of the sea in the opposite. There is evidently an excess of heavy material vertically beneath the middle of the Pacific; while not very remote from the point of the globe diametrically opposite rises the great table-land of India and the Himalaya chain, on the summits of which the air has not more than a third of the density it has on the sea-level, and from which animated existence is for ever excluded.—*Herschel's Outlines*, 5th edit.

LIGHT OF THE MOON.

The actual illumination of the lunar surface is not much superior to that of weathered sandstone-rock in full sunshine. Sir John Herschel has frequently compared the moon setting behind the gray perpendicular façade of the Table Mountain at the Cape of Good Hope, illuminated by the sun just risen from the opposite quarter of the horizon, when it has been scarcely distinguishable in brightness from the rock in contact with it. The sun and moon being nearly at equal altitudes, and the atmosphere perfectly free from cloud or vapour, its effect is alike on both luminaries.

HEAT OF MOONLIGHT.

M. Zantedeschi has proved, by a long series of experiments in the Botanic Gardens at Venice, Florence, and Padua, that, contrary to the general opinion, the diffused rays of moonlight have an influence upon the organs of plants, as the Sensitive Plant and the *Desmodium gyrans*. The influence was feeble compared with that of the sun; but the action is left beyond further question.

Melloni has proved that the rays of the Moon give out a slight degree of Heat (see *Things not generally Known*, p. 7); and Professor Piazzi Smyth, from a point of the Peak of Teneriffe 8840 feet above the sea-level, has found distinctly perceptible the heat radiated from the moon, which has been so often sought for in vain in a lower region.

SCENERY OF THE MOON.

By means of the telescope, mountain-peaks are distinguished in the ash-gray light of the larger spots and isolated brightly-shining points of the moon, even when the disc is already more than half illuminated. Lambert and Schroter have shown that the extremely variable intensity of the ash-gray light of the moon depends upon the greater or less degree of reflection of the sunlight which falls upon the earth, according as it is reflected from continuous continental masses, full of sandy deserts, grassy steppes, tropical forests, and barren rocky ground, or from large ocean surfaces. Lambert made the remarkable observation (14th of February 1774) of a change of the ash-coloured moonlight into an olive-green colour bordering upon yellow. "The moon, which then stood vertically over the Atlantic Ocean, received upon its right side the green terrestrial light which is reflected towards her when the sky is clear by the forest districts of South America."

Plutarch says distinctly, in his remarkable work *On the Face in the Moon*, that we may suppose the *spots* to be partly deep chasms and valleys, partly mountain-peaks, which cast long shadows, like Mount Athos, whose shadow reaches Lemnos. The spots cover about two-fifths of the whole disc. In a clear atmosphere, and under favourable circumstances in the position of the moon, some of the spots are visible to the naked eye; as the edge of the Apennines, the dark elevated plain Grimaldus, the enclosed *Mare Crisium*, and Tycho, crowded round with numerous mountain ridges and craters.

Professor Alexander remarks, that a map of the eastern hemisphere, taken with the Bay of Bengal in the centre, would bear a striking resemblance to the face of the moon presented to us. The dark portions of the moon he considers to be continental elevations, as shown by measuring the average height of mountains above the dark and the light portions of the moon.

The surface of the moon can be as distinctly seen by a good telescope magnifying 1000 times, as it would be if not more than 250 miles distant.

LIFE IN THE MOON.

A circle of one second in diameter, as seen from the earth, on the surface of the moon contains about a square mile. Telescopes, therefore, must be greatly improved before we could expect to see signs of inhabitants, as manifested by edifices or changes on the surface of the soil. It should, however, be observed, that owing to the small density of the materials of the moon, and the comparatively feeble gravitation of bodies on her surface, muscular force would there go six times as far in overcoming the weight of materials as on the earth. Owing to the want of air, however, it seems impossible that any form of life analogous to those on earth can subsist there. No appearance indicating vegetation, or the slightest variation of surface which can in our opinion fairly be ascribed to change of season, can any where be discerned.—*Sir John Herschel's Outlines.*

THE MOON SEEN THROUGH LORD ROSSE'S TELESCOPE.

In 1846, the Rev. Dr. Scoresby had the gratification of observing the Moon through the stupendous telescope constructed by Lord Rosse at Parsonstown. It appeared like a globe of molten silver, and every object to the extent of 100 yards was quite visible. Edifices, therefore, of the size of York Minster, or even of the ruins of Whitby Abbey, might be easily perceived, if they had existed. But there was no appearance of any thing of that nature; neither was there any indication of the existence of water, or of an atmosphere. There were a great number of extinct volcanoes, several miles in breadth; through one of them there was a line of continuance about 150 miles in length, which ran in a straight direction, like a railway. The general appearance, however, was like one vast ruin of nature; and many of the pieces of rock driven out of the volcanoes appeared to lie at various distances.

MOUNTAINS IN THE MOON.

By the aid of telescopes, we discern irregularities in the surface of the moon which can be no other than mountains and valleys,—for this plain reason, that we see the shadows cast by the former in the exact proportion as to length which they ought to have when we take into account the inclinations of the sun's rays to that part of the moon's surface on which they stand. From micrometrical measurements of the lengths of the shadows of the more conspicuous mountains, Messrs. Baer and Maedler have given a list of heights for no less than 1095 lunar mountains, among which occur all degrees of elevation up to 22,823 British feet, or about 1400 feet higher than Chimborazo in the Andes.

If Chimborazo were as high in proportion to the earth's diameter as a mountain in the moon known by the name of Newton is to the moon's diameter, its peak would be more than sixteen miles high.

Arago calls to mind, that with a 6000-fold magnifying power, which nevertheless could not be applied to the moon with proportionate results, the mountains upon the moon would appear to us just as Mont Blanc does to the naked eye when seen from the Lake of Geneva.

We sometimes observe more than half the surface of the moon, the eastern and northern edges being more visible at one time, and the western or southern at another. By means of this libration we are enabled to see the annular mountain Malapert (which occasionally conceals the moon's south pole), the arctic landscape round the crater of Gioja, and the large gray plane near Endymion, which conceals in superficial extent the *mare vaporum*.

Three-sevenths of the moon are entirely concealed from our observation; and must always remain so, unless some new and unexpected disturbing causes come into play.—*Humboldt.*

The first object to which Galileo directed his telescope was the mountainous parts of the moon, when he showed how their summits might be measured: he found in the moon some circular districts surrounded on all sides by mountains similar to the form of Bohemia. The measurements of the mountains were made by the method of the tangents of the solar ray. Galileo, as Helvetius did still later, measured the distance of the summit of the mountains from the boundary of the illuminated portion at the moment when the mountain summit was first struck by the solar ray. Humboldt found no observations of the lengths of the shadows of the mountains: the summits were "much higher than the mountains on our earth." The comparison is remarkable, since, according to Riccioli, very exaggerated ideas of the height of our mountains were then entertained. Galileo like all other observers up to the close of the eighteenth century, believed in the existence of many seas and of a lunar atmosphere.

THE MOON AND THE WEATHER.

The only influence of the Moon on the Weather of which we have any decisive evidence is the tendency to disappearance of clouds under the full moon, which Sir John Herschel refers to its heat being much more readily absorbed in traversing transparent media than direct solar heat, and being extinguished in the upper regions of our atmosphere, never reaches the surface of the atmosphere at all.

THE MOON'S ATTRACTION.

Mr. G. P. Bond of Cambridge, by some investigations to ascertain whether the Attraction of the Moon has any effect upon the motion of a pendulum, and consequently upon the rate of a clock, has found the last to be changed to the amount of 9/1000 of a second daily. At the equator the moon's attraction changes the weight of a body only 1/7000000 of the whole; yet this force is sufficient to produce the vast phenomena of the tides!

It is no slight evidence of the importance of analysis, that Laplace's perfect theory of tides has enabled us in our astronomical ephemerides to predict the height of spring-tides at the periods of new and full moon, and thus put the inhabitants of the sea on their guard against the increased danger attending the lunar revolutions.

MEASURING THE EARTH BY THE MOON.

As the form of the Earth exerts a powerful influence on the motion of other cosmical bodies, and especially on that of its neighbouring satellite, a more perfect knowledge of the motion of the latter will enable us reciprocally to draw an inference regarding the figure of the earth. Thus, as Laplace ably remarks: "an astronomer, without leaving his observatory, may, by a comparison of lunar theory with true observations, not only be enabled to determine the form and size of the earth, but also its distance from the sun and moon; results that otherwise could only be arrived at by long and arduous expeditions to the most remote parts of both hemispheres." The compression which may be inferred from lunar inequalities affords an advantage not yielded by individual measurements of degrees or experiments with the pendulum, since it gives a mean amount which is referable to the whole planet.—*Humboldt's Cosmos*, vol. i.

The distance of the moon from the earth is about 240,000 miles; and if a railway-carriage were to travel at the rate of 1000 miles a-day, it would be eight months in reaching the moon. But that is nothing compared with the length of time it would occupy a locomotive to reach the sun from the earth: if travelling at the rate of 1000 miles a-day, it would require 260 years to reach it.

CAUSE OF ECLIPSES.

As the Moon is at a very moderate distance from us (astronomically speaking), and is in fact our nearest neighbour, while the sun and stars are in comparison immensely beyond it, it must of necessity happen that at one time or other it must *pass over*, and *occult* or *eclipse*, every star or planet within its zone, and, as seen from the *surface* of the earth, even somewhat beyond it. Nor is the sun itself exempt from being thus hidden whenever any part of the moon's disc, in this her tortuous course, comes to *overlap* any part of the space occupied in the heavens by that luminary. On these occasions is exhibited the most striking and impressive of all the occasional phenomena of astronomy, an *Eclipse of the Sun*, in which a greater or less portion, or even in some conjunctures the whole of its disc, is obscured, and, as it were, obliterated, by the superposition of that of the moon, which appears upon it as a circularly-terminated black spot, producing a temporary diminution of daylight, or even nocturnal darkness, so that the stars appear as if at midnight.—*Sir John Herschel's Outlines.*

VAST NUMBERS IN THE UNIVERSE.

The number of telescopic stars in the Milky Way uninterrupted by any nebulæ is estimated at 18,000,000. To compare this number with something analogous, Humboldt calls attention to the fact, that there are not in the whole heavens more than about 8000 stars, between the first and the sixth magnitudes, visible to the naked eye. The barren astonishment excited by numbers and dimensions in space when not considered with reference to applications engaging the mental and perceptive powers of man, is awakened in both extremes of the universe—in the celestial bodies as in the minutest animalcules. A cubic inch of the polishing slate of Bilin contains, according to Ehrenberg, 40,000 millions of the siliceous shells of Galionellæ.

FOR WHAT PURPOSE WERE THE STARS CREATED?

Surely not (says Sir John Herschel) to illuminate *our* nights, which an additional moon of the thousandth part of the size of our own would do much better; nor to sparkle as a pageant void of meaning and reality, and bewilder us among vain conjectures. Useful, it is true, they are to man as points of exact and permanent reference; but he must have studied astronomy to little purpose, who can suppose man to be the only object of his Creator's care, or who does not see in the vast and wonderful apparatus around us provision for other races of animated beings. The planets derive their light from the sun; but that cannot be the case with the stars. These doubtless, then, are themselves suns; and may perhaps, each in its sphere, be the presiding centre round which other planets, or bodies of which we can form no conception from any analogy offered by our own system, are circulating.19

NUMBER OF STARS.

Various estimates have been hazarded on the Number of Stars throughout the whole heavens visible to us by the aid of our colossal telescopes. Struve assumes for Herschel's 20-feet reflector, that a magnifying power of 180 would give 5,800,000 for the number of stars lying within the zones extending 30° on either side of the equator, and 20,374,000 for the whole heavens. Sir William Herschel conjectured that 18,000,000 of stars in the Milky Way might be seen by his still more powerful 40-feet reflecting telescope.—*Humboldt's Cosmos*, vol. iii.

The assumption that the extent of the starry firmament is literally infinite has been made by Dr. Olbers the basis of a conclusion that the celestial spaces are in some slight degree deficient in *transparency*; so that all beyond a certain distance is and must remain for ever unseen, the geometrical progression of the extinction of light far outrunning the effect of any conceivable increase in the power of our telescopes. Were it not so, it is argued that every part of the celestial concave ought to shine with the brightness of the solar disc, since no visual ray could be so directed as not, in some point or other of its infinite length, to encounter such a disc.—*Edinburgh Review*, Jan. 1848.

STARS THAT HAVE DISAPPEARED.

Notwithstanding the great accuracy of the catalogued positions of telescopic fixed stars and of modern star-maps, the certainty of conviction that a star in the heavens has actually disappeared since a certain epoch can only be arrived at with great caution. Errors of actual observation, of reduction, and of the press, often disfigure the very best catalogues. The disappearance of a heavenly body from the place in which it had been before distinctly seen, may be the result of its own motion as much as of any such diminution of its photometric process as would render the waves of light too weak to excite our organs of sight. What we no longer see, is not necessarily annihilated. The idea of destruction or combustion, as applied to disappearing stars, belongs to the age of Tycho Brahe. Even Pliny makes it a question. The apparent eternal cosmical alternation of existence and destruction is not annihilation; it is merely the transition of matter into new forms, into combinations which are subject to new processes. Dark cosmical bodies may by a renewed process of light again become luminous.—*Humboldt's Cosmos*, vol. iii.

THE POLE-STAR FOUR THOUSAND YEARS AGO.

Sir John Herschel, in his *Outlines of Astronomy*, thus shows the changes in the celestial pole in 4000 years:

At the date of the erection of the Pyramid of Gizeh, which precedes the present epoch by nearly 4000 years, the longitudes of all the stars were less by 55° 45′ than at present. Calculating from this datum the place of the pole of the heavens among the stars, it will be found to fall near α Draconis; its distance from that star being 3° 44′ 25″. This being the most conspicuous star in the immediate neighbourhood, was therefore the Pole Star of that epoch. The latitude of Gizeh being just 30° north, and consequently the altitude of the North Pole there also 30°, it follows that the star in question must have had at its lowest culmination at Gizeh an altitude of 25° 15' 35". Now it is a remarkable fact, that of the nine pyramids still existing at Gizeh, six (including all the largest) have the narrow passages by which alone they can be entered (all which open out on the northern faces of their respective pyramids) inclined to the horizon downwards at angles the mean of which is 26° 47'. At the bottom of every one of these passages, therefore, the Pole Star must have been visible at its lower culmination; a circumstance which can hardly be supposed to have been unintentional, and was doubtless connected (perhaps superstitiously) with the astronomical observations of that star, of whose proximity to the pole at the epoch of the erection of these wonderful structures we are thus furnished with a monumental record of the most imperishable nature.

THE PLEIADES.

The Pleiades prove that, several thousand years ago even as now, stars of the seventh magnitude were invisible to the naked eye of average visual power. The group consists of seven stars, of which six only, of the third, fourth, and fifth magnitudes, could be readily distinguished. Of these Ovid says (*Fast.* iv. 170):

"Quæ septem dici, sex tamen esse solent."

Aratus states there were only six stars visible in the Pleiades.

One of the daughters of Atlas, Merope, the only one who was wedded to a mortal, was said to have veiled herself for very shame and to have disappeared. This is probably the star of the seventh magnitude, which we call Celæne; for Hipparchus, in his commentary on Aratus, observes that on clear moonless nights *seven stars* may actually be seen.

The Pleiades were doubtless known to the rudest nations from the earliest times; they are also called the *mariner's stars*. The name is from πλεῖν (*plein*), 'to sail.' The navigation of the Mediterranean lasted from May to the beginning of November, from the early rising to the early setting of the Pleiades. In how many beautiful effusions of poetry and sentiment has "the Lost Pleiad" been deplored!—and, to descend to more familiar illustration of this group, the "Seven Stars," the sailors' favourites, and a frequent river-side public-house sign, may be traced to the Pleiades.

CHANGE OF COLOUR IN THE STARS.

The scintillation or twinkling of the stars is accompanied by variations of colour, which have been remarked from a very early age. M. Arago states, upon the authority of M. Babinet, that the name of Barakesch, given by the Arabians to Sirius, signifies *the star of a thousand colours*; and Tycho Brahe, Kepler, and others, attest to similar change of colour in twinkling. Even soon after the invention of the telescope, Simon Marius remarked that by removing the eye-piece of the telescope the images of the stars exhibited rapid fluctuations of brightness and colour. In 1814 Nicholson applied to the telescope a smart vibration, which caused the image of the star to be transformed into a curved line of light returning into itself, and diversified by several colours; each colour occupied about a third of the whole length of the curve, and by applying ten vibrations in a second, the light of Sirius in that time passed through thirty changes of colour. Hence the stars in general shine only by a portion of their light, the effect of twinkling being to diminish their brightness. This phenomenon M. Arago explains by the principle of the interference of light.

Ptolemy is said to have noted Sirius as a *red* star, though it is now white. Sirius twinkles with red and blue light, and Ptolemy's eyes, like those of several other persons, may have been more sensitive to the *red* than to the *blue* rays.—*Sir David Brewster's More Worlds than One*, p. 235.

Some of the double stars are of very different and dissimilar colours; and to the revolving planetary bodies which apparently circulate around them, a day lightened by a red light is succeeded by, not a night, but a day equally brilliant, though illuminated only by a green light.

DISTANCE OF THE NEAREST FIXED STAR FROM THE EARTH.

Sir John Herschel wrote in 1833: "What is the distance of the nearest fixed star? What is the scale on which our visible firmament is constructed? And what proportion do its dimensions bear to those of our own immediate system? To this, however, astronomy has hitherto proved unable to supply an answer. All we know on this subject is negative." To these questions, however, an answer can now be given. Slight changes of position of some of the stars, called parallax, have been distinctly observed and measured; and among these stars No. 61 Cygni of Flamstead's catalogue has a parallax of 5″, and that of α Centauri has a proper motion of 4″ per annum.

The same astronomer states that each second of parallax indicates a distance of 20 billions of miles, or 3¼ years' journey of light. Now the light sent to us by the sun, as compared with that sent by Sirius and α Centauri, is about 22 thousand millions to 1. "Hence, from the parallax assigned above to that star, it is easy to conclude that its intrinsic splendour, as compared with that of our sun at equal distances, is 2•3247, that of the sun being unity. The light of Sirius is four times that of α Centauri, and its parallax only 0•15″. This, in effect, ascribes to it an intrinsic splendour equal to 96•63 times that of α Centauri, and therefore 224•7 times that of our sun."

This is justly regarded as one of the most brilliant triumphs of astronomical science, for the delicacy of the investigation is almost inconceivable; yet the reasoning is as unimpeachable as the demonstration of a theorem of Euclid.

LIGHT OF A STAR SIXTEENFOLD THAT OF THE SUN.

The bright star in the constellation of the Lyre, termed Vega, is the brightest in the northern hemisphere; and the combined researches of Struve, father and son, have found that the distance of this star from the earth is no less than 130 billions of miles! Light travelling at the rate of 192 thousand miles in a second consequently occupies twenty-one years in passing from this star to the earth. Now it has been found, by comparing the light of Vega with the light of the sun, that if the latter were removed to the distance of 130 billions of miles, his apparent brightness would not amount to more than the sixteenth part of the apparent brightness of Vega. We are therefore warranted in concluding that the light of Vega is equal to that of sixteen suns.

DIVERSITIES OF THE PLANETS.

In illustration of the great diversity of the physical peculiarities and probable condition of the planets, Sir John Herschel describes the intensity of solar radiation as nearly seven times greater on Mercury than on the earth, and on Uranus 330 times less; the proportion between the two extremes being that of upwards of 2000 to 1. Let any one figure to himself, (adds Sir John,) the condition of our globe were the sun to be septupled, to say nothing of the greater ratio; or were it diminished to a seventh, or to a 300th of its actual power! Again, the intensity of gravity, or its efficacy in counteracting muscular power and repressing animal activity, on Jupiter is nearly two-and-a-half times that on the earth; on Mars not more than one-half; on the moon one-sixth; and on the smaller planets probably not more than one-twentieth; giving a scale of which the extremes are in the proportion of sixty to one. Lastly, the density of Saturn hardly exceeds one-eighth of the mean density of the earth, so that it must consist of materials not much heavier than cork.

Jupiter is eleven times, Saturn ten times, Uranus five times, and Neptune nearly six times, the diameter of our earth.

These four bodies revolve in space at such distances from the sun, that if it were possible to start thence for each in succession, and to travel at the railway speed of 33 miles per hour, the traveller would reach

Jupiter in   1712 years

Saturn   3113   "

Uranus   6226   "

Neptune   9685   "

If, therefore, a person had commenced his journey at the period of the Christian era, he would now have to travel nearly 1300 years before he would arrive at the planet Saturn; more than 4300 years before he would reach Uranus; and no less than 7800 years before he could reach the orbit of Neptune.

Yet the light which comes to us from these remote confines of the solar system first issued from the sun, and is then reflected from the surface of the planet. When the telescope is turned towards Neptune, the observer's eye sees the object by means of light that issued from the sun eight hours before, and which since then has passed nearly twice through that vast space which railway speed would require almost a century of centuries to accomplish.—*Bouvier's Familiar Astronomy.*

GRAND RESULTS OF THE DISCOVERY OF JUPITER'S SATELLITES.

This discovery, one of the first fruits of the invention of the telescope, and of Galileo's early and happy idea of directing its newly-found powers to the examination of the heavens, forms one of the most memorable epochs in the history of astronomy. The first astronomical solution of the great problem of *the longitude*, practically the most important for the interests of mankind which has ever been brought under the dominion of strict scientific principles, dates immediately from this discovery. The final and conclusive establishment of the Copernican system of astronomy may also be considered as referable to the discovery and study of this exquisite miniature system, in which the laws of the planetary motions, as ascertained by Kepler, and specially that which connects their periods and distances, were specially traced, and found to be satisfactorily maintained. And (as if to accumulate historical interest on this point) it is to the observation of the eclipses of Jupiter's satellites that we owe the grand discovery of the aberration of light, and the consequent determination of the enormous velocity of that wonderful element—192,000 miles per second. Mr. Dawes, in 1849, first noticed the existence of round, well-defined, bright spots on the belts of Jupiter. They vary in situation and number, as many as ten having been seen on one occasion. As the belts of Jupiter have been ascribed to the existence of currents analogous to our trade-winds, causing the body of Jupiter to be visible through his cloudy atmosphere, Sir John Herschel conjectures that those bright spots may possibly be insulated masses of clouds of local origin, similar to the cumuli which sometimes cap ascending columns of vapour in our atmosphere.

It would require nearly 1300 globes of the size of our earth to make one of the bulk of Jupiter. A railway-engine travelling at the rate of thirty-three miles an hour would travel round the earth in a month, but would require more than eleven months to perform a journey round Jupiter.

WAS SATURN'S RING KNOWN TO THE ANCIENTS?

In Maurice's *Indian Antiquities* is an engraving of Sani, the Saturn of the Hindoos, taken from an image in a very ancient pagoda, which represents the deity encompassed by a *ring* formed of two serpents. Hence it is inferred that the ancients were acquainted with the existence of the ring of Saturn.

Arago mentions the remarkable fact of the ring and fourth satellite of Saturn having been seen by Sir W. Herschel with his smaller telescope by the naked eye, without any eye-piece.

The first or innermost of Saturn's satellites is nearer to the central body than any other of the secondary planets. Its distance from the centre of Saturn is 80,088 miles; from the surface of the planet 47,480 miles; and from the outmost edge of the ring only 4916 miles. The traveller may form to himself an estimate of the smallness of this amount by remembering the statement of the well-known navigator, Captain Beechey, that he had in three years passed over 72,800 miles.

According to very recent observations, Saturn's ring is divided into *three* separate rings, which, from the calculations of Mr. Bond, an American astronomer, must be fluid. He is of opinion that the number of rings is continually changing, and that their maximum number, in the normal condition of the mass, does not exceed *twenty.* Mr. Bond likewise maintains that the power which sustains the centre of gravity of the *ring* is not in the planet itself, but in its satellites; and the satellites, though constantly disturbing the ring, actually sustain it in the very act of perturbation. M. Otto Struve and Mr. Bond have lately studied with the great Munich telescope, at the observatory of Pulkowa, the *third* ring of Saturn, which Mr. Lassell and Mr. Bond discovered to be *fluid.* They saw distinctly the dark interval between this fluid ring and the two old ones, and even measured its dimensions; and they perceived at its inner margin an edge feebly illuminated, which they thought might be the commencement of a fourth ring. These astronomers are of opinion, that the fluid ring is not of very recent formation, and that it is not subject to rapid change; and they have come to the extraordinary conclusion, that the inner border of the ring has, since the time of Huygens, been gradually approaching to the body of Saturn, and that *we may expect, sooner or later, perhaps in some dozen of years, to see the rings united with the body of the planet.* But this theory is by other observers pronounced untenable.

TEMPERATURE OF THE PLANET MERCURY.

Mercury being so much nearer to the Sun than the Earth, he receives, it is supposed, seven times more heat than the earth. Mrs. Somerville says: "On Mercury, the mean heat arising from the intensity of the sun's rays must be above that of boiling quicksilver, and water would boil even at the poles." But he may be provided with an atmosphere so constituted as to absorb or reflect a great portion of the superabundant heat; so that his inhabitants (if he have any) may enjoy a climate as temperate as any on our globe.

SPECULATIONS ON VESTA AND PALLAS.

The most remarkable peculiarities of these ultra-zodiacal planets, according to Sir John Herschel, must lie in this condition of their state: a man placed on one of them would spring with ease sixty feet high, and sustain no greater shock in his descent than he does on the earth from leaping a yard. On such planets, giants might exist; and those enormous animals which on the earth require the buoyant power of water to counteract their weight, might there be denizens of the land. But of such speculations there is no end.

IS THE PLANET MARS INHABITED?

The opponents of the doctrine of the Plurality of Worlds allow that a greater probability exists of Mars being inhabited than in the case of any other planet. His diameter is 4100 miles; and his surface exhibits spots of different hues,—the *seas*, according to Sir John Herschel, being *green*, and the land *red*. "The variety in the spots," says this astronomer, "may arise from the planet not being destitute of atmosphere and cloud; and what adds greatly to the probability of this, is the appearance of brilliant white spots at its poles, which have been conjectured, with some probability, to be snow, as they disappear when they have been long exposed to the sun, and are greatest when emerging from the long night of their polar winter, the snow-line then extending to about six degrees from the pole." "The length of the day," says Sir David Brewster, "is almost exactly twenty-four hours,—the same as that of the earth. Continents and oceans and green savannahs have been observed upon Mars, and the snow of his polar regions has been seen to disappear with the heat of summer." We actually see the clouds floating in the atmosphere of Mars, and there is the appearance of land and water on his disc. In a sketch of this planet, as seen in the pure atmosphere of Calcutta by Mr. Grant, it appears, to use his words, "actually as a little world," and as the earth would appear at a distance, with its seas and continents of different shades. As the diameter of Mars is only about one half that of our earth, the weight of bodies will be about one half what it would be if they were placed upon our globe.

DISCOVERY OF THE PLANET NEPTUNE.

This noble discovery marked in a signal manner the maturity of astronomical science. The proof, or at least the urgent presumption, of the existence of such a planet, as a means of accounting (by its attraction) for certain small irregularities observed in the motions of Uranus, was afforded almost simultaneously by the independent researches of two geometers, Mr. Adams of Cambridge, and M. Leverrier of Paris, who were enabled *from theory alone* to calculate whereabouts it ought to appear in the heavens, *if visible*, the places thus independently calculated agreeing surprisingly.*Within a single degree* of the place assigned by M. Leverrier's calculations, and by him communicated to Dr. Galle of the Royal Observatory at Berlin, it was actually found by that astronomer on the very first night after the receipt of that communication, on turning a telescope on the spot, and comparing the stars in its immediate neighbourhood with those previously laid down in one of the zodiacal charts. This remarkable verification of an indication so extraordinary took place on the 23d of September 1846.[20]—*Sir John Herschel's Outlines.*

Neptune revolves round the sun in about 172 years, at a mean distance of thirty,—that of Uranus being nineteen, and that of the earth one: and by its discovery the solar system has been extended *one thousand millions of miles* beyond its former limit.

Neptune is suspected to have a ring, but the suspicion has not been confirmed. It has been demonstrated by the observations of Mr. Lassell, M. Otto Struve, and Mr. Bond, to be attended by at least one satellite.

One of the most curious facts brought to light by the discovery of Neptune, is the failure of Bode's law to give an approximation to its distance from the sun; a striking exemplification of the danger of trusting to the universal applicability of an empirical law. After standing the severe test which led to the discovery of the asteroids, it seemed almost contrary to the laws of probability that the discovery of another member of the planetary system should prove its failure as an universal rule.

MAGNITUDE OF COMETS.

Although Comets have a smaller mass than any other cosmical bodies—being, according to our present knowledge, probably not equal to 1/5000th part of the earth's mass—yet they occupy the largest space, as their tails in several instances extend over many millions of miles. The cone of luminous vapour which radiates from them has been found in some cases (as in 1680 and 1811) equal to the length of the earth's distance from the sun, forming a line that intersects both the orbits of Venus and

Mercury. It is even probable that the vapour of the tails of comets mingled with our atmosphere in the years 1819 and 1823.—*Humboldt's Cosmos*, vol. i.

COMETS VISIBLE IN SUNSHINE—THE GREAT COMET OF 1843.

The phenomenon of the tail of a Comet being visible in bright Sunshine, which is recorded of the comet of 1402, occurred again in the case of the large comet of 1843, whose nucleus and tail were seen in North America on February 28th (according to the testimony of J. G. Clarke, of Portland, State of Maine), between one and three o'clock in the afternoon. The distance of the very dense nucleus from the sun's light admitted of being measured with much exactness. The nucleus and tail (a darker space intervening) appeared like a very pure white cloud.—*American Journal of Science*, vol. xiv.

E. C. Otté, the translator of Bohn's edition of Humboldt's *Cosmos*, at New Bedford, Massachusetts, U.S., Feb. 28th, 1843, distinctly saw the above comet between one and two in the afternoon. The sky at the time was intensely blue, and the sun shining with a dazzling brightness unknown in European climates.

This very remarkable Comet, seen in England on the 17th of March 1843, had a nucleus with the appearance of a planetary disc, and the brightness of a star of the first or second magnitude. It had a double tail divided by a dark line. At the Cape of Good Hope it was seen in full daylight, and in the immediate vicinity of the sea; but the most remarkable fact in its history was its near approach to the sun, its distance from his surface being only *one-fourteenth* of his diameter. The heat to which it was exposed, therefore, was much greater than that which Sir Isaac Newton ascribed to the comet of 1680, namely 200 times that of red-hot iron. Sir John Herschel has computed that it must have been 24 times greater than that which was produced in the focus of Parker's burning lens, 32 inches in diameter, which melts crystals of quartz and agate.[21]

THE MILKY WAY UNFATHOMABLE.

M. Struve of Pulkowa has compared Sir William Herschel's opinion on this subject, as maintained in 1785, with that to which he was subsequently led; and arrives at the conclusion that, according to Sir W. Herschel himself, the visible extent of the Milky Way increases with the penetrating power of the telescopes employed; that it is impossible to discover by his instruments the termination of the Milky Way (as an independent cluster of stars); and that even his gigantic telescope of forty feet focal length does not enable him to extend our knowledge of the Milky Way, which is incapable of being sounded. Sir William Herschel's *Theory of the Milky Way* was as follows: He considered our solar system, and all the stars which we can see with the eye, as placed within, and constituting a part of, the nebula of the Milky Way, a congeries of many millions of stars, so that the projection of these stars must form a luminous track on the concavity of the sky; and by estimating or counting the number of stars in different directions, he was able to form a rude judgment of the probable form of the nebula, and of the probable position of the solar system within it.

This remarkable belt has maintained from the earliest ages the same relative situation among the stars; and, when examined through powerful telescopes, is found (wonderful to relate!) *to consist entirely of stars scattered by millions,* like glittering dust, on the black ground of the general heavens.

DISTANCES OF NEBULÆ.

These are truly astounding. Sir William Herschel estimated the distance of the annular nebula between Beta and Gamma Lyræ to be from our system 950 times that of Sirius; and a globular cluster about 5½° south-east of Beta Sir William computed to be one thousand three hundred billions of miles from our system. Again, in Scutum Sobieski is one nebula in the shape of a horseshoe; but which, when viewed with high magnifying power, presents a different appearance. Sir William Herschel estimated this nebula to be 900 times farther from us than Sirius. In some parts of its vicinity he observed 588 stars in his telescope at one time; and he counted 258,000 in a space 10° long and 2½° wide. There is a globular cluster between the mouths of Pegasus and Equuleus, which Sir William Herschel estimated to be 243 times farther from us than Sirius. Caroline Herschel discovered in the right foot of Andromeda a nebula of enormous dimensions, placed at an inconceivable distance from us: it consists probably of myriads of solar systems, which, taken together, are but a point in the universe. The nebula about 10° west of the principal star in Triangulum is supposed by Sir William Herschel to be 344 times the distance of Sirius from the earth, which would be the immense sum of nearly seventeen thousand billions of miles from our planet.

INFINITE SPACE.

After the straining mind has exhausted all its resources in attempting to fathom the distance of the smallest telescopic star, or the faintest nebula, it has reached only the visible confines of the sidereal creation. The universe of stars is but an atom in the universe of space; above it, and beneath it, and around it, there is still infinity.

The commencement of our Planetary System, including the sun, must, according to Kant and Laplace, be regarded as an immense nebulous mass filling the portion of space which is now occupied by our system far beyond the limits of Neptune, our most distant planet. Even now we perhaps see similar masses in the distant regions of the firmament, as patches of nebulæ, and nebulous stars; within our system also, comets, the zodiacal light, the corona of the sun during a total eclipse, exhibit resemblances of a nebulous substance, which is so thin that the light of the stars passes through it unenfeebled and unrefracted. If we calculate the density of the mass of our planetary system, according to the above assumption, for the time when it was a nebulous sphere which reached to the path of the outmost planet, we should find that it would require several cubic miles of such matter to weigh a single grain.— *Professor Helmholtz.*

A quarter of a century ago, Sir John Herschel expressed his opinion that those nebulæ which were not resolved into individual stars by the highest powers then used, might be hereafter completely resolved by a further increase of optical power:

In fact, this probability has almost been converted into a certainty by the magnificent reflecting telescope constructed by Lord Rosse, of 6 feet in aperture, which has resolved, or rendered resolvable, multitudes of nebulæ which had resisted all inferior powers. The sublimity of the spectacle afforded by that instrument of some of the larger globular and other clusters is declared by all who have witnessed it to be such as no words can express.[23]

Although, therefore, nebulæ do exist, which even in this powerful telescope appear as nebulæ, without any sign of resolution, it may very reasonably be doubted whether there be really any essential physical distinction between nebulæ and clusters of stars, at least in the nature of the matter of which they consist; and whether the distinction between such nebulæ as are easily resolved, barely resolvable with excellent telescopes, and altogether irresolvable with the best, be any thing else than one of degree, arising merely from the excessive minuteness and multitude of the stars of which the latter, as compared with the former, consist.—*Outlines of Astronomy*, 5th edit. 1858.

It should be added, that Sir John Herschel considers the "nebular hypothesis" and the above theory of sidereal aggregation to stand quite independent of each other.

ORIGIN OF HEAT IN OUR SYSTEM.

Professor Helmholtz, assuming that at the commencement the density of the nebulous matter was a vanishing quantity, as compared with the present density of the sun and planets, calculates how much work has been performed by the condensation; how much of this work still exists in the form of mechanical force, as attraction of the planets towards the sun, and as *vis viva* of their motion; and finds by this how much of the force has been converted into heat.

The result of this calculation is, that only about the 45th part of the original mechanical force remains as such, and that the remainder, converted into heat, would be sufficient to raise a mass of water equal to the sun and planets taken together, not less than 28,000,000 of degrees of the centigrade scale. For the sake of comparison, Professor Helmholtz mentions that the highest temperature which we can produce by the oxy-hydrogen blowpipe, which is sufficient to vaporise even platina, and which but few bodies can endure, is estimated at about 2000 degrees. Of the action of a temperature of 28,000,000 of such degrees we can form no notion. If the mass of our entire system were of pure coal, by the combustion of the whole of it only the 350th part of the above quantity would be generated.

The store of force at present possessed by our system is equivalent to immense quantities of heat. If our earth were by a sudden shock brought to rest in her orbit—which is not to be feared in the existing arrangement of our system—by such a shock a quantity of heat would be generated equal to that produced by the combustion of fourteen such earths of solid coal. Making the most unfavourable assumption as to its capacity for heat, that is, placing it equal to that of water, the mass of the earth would thereby be heated 11,200°; it would therefore be quite fused, and for the most part reduced to vapour. If, then, the earth, after having been thus brought to rest, should fall into the sun, which of course would be the case, the quantity of heat developed by the shock would be 400 times greater.

AN ASTRONOMER'S DREAM VERIFIED.

The most fertile region in astronomical discovery during the last quarter of a century has been the planetary members of the solar system. In 1833, Sir John Herschel enumerated ten planets as visible from the earth, either by the unaided eye or by the telescope; the number is now increased more than fivefold. With the exception of Neptune, the discovery of new planets is confined to the class called Asteroids. These all revolve in elliptic orbits between those of Jupiter and Mars. Zitius of Wittemberg discovered an empirical law, which seemed to govern the distances of the planets from the sun; but there was a remarkable interruption in the law, according to which a planet ought to have been placed between Mars and Jupiter. Professor Bode of Berlin directed the attention of astronomers to the possibility of such a planet existing; and in seven years' observations from the commencement of the present century, not one but four planets were found, differing widely from one another in the elements of their orbits, but agreeing very nearly at their mean distances from the sun with that of the supposed planet. This curious coincidence of the mean distances of these four asteroids with the planet according to Bode's law, as it is generally called, led to the conjecture that these four planets were but fragments of the missing planet, blown to atoms by some internal explosion, and that many more fragments might exist, and be possibly discovered by diligent search.

Concerning this apparently wild hypothesis, Sir John Herschel offered the following remarkable apology: "This may serve as a specimen of the dreams in which astronomers, like other speculators, occasionally and harmlessly indulge."

The dream, wild as it appeared, has been realised now. Sir John, in the fifth edition of his *Outlines of Astronomy*, published in 1858, tells us:

Whatever may be thought of such a speculation as a physical hypothesis, this conclusion has been verified to a considerable extent as a matter of fact by subsequent discovery, the result of a careful and minute examination and mapping down of the smaller stars in and near the zodiac, undertaken with that express object. Zodiacal charts of this kind, the product of the zeal and industry of many astronomers, have been constructed, in which every star down to the ninth, tenth, or even lower magnitudes, is inserted; and these stars being compared with the actual stars of the heavens, the intrusion of any stranger within their limits cannot fail to be noticed when the comparison is systematically conducted. The discovery of Astræa and Hebe by Professor Hencke, in 1845 and 1847, revived the flagging spirit of inquiry in this direction; with what success, the list of fifty-two asteroids, with their names and the dates of their discovery, will best show. The labours of our indefatigable countryman, Mr. Hind, have been rewarded by the discovery of no less than eight of them.

FIRE-BALLS AND SHOOTING STARS.

Humboldt relates, that a friend at Popayan, at an elevation of 5583 feet above the sea-level, at noon, when the sun was shining brightly in a cloudless sky, saw his room lighted up by a fire-ball: he had his back towards the window at the time, and on turning round, perceived that great part of the path traversed by the fire-ball was still illuminated by the brightest radiance. The Germans call these phenomena *star-snuff*, from the vulgar notion that the lights in the firmament undergo a process of snuffing, or cleaning. Other nations call it *a shot or fall of stars*, and the English *star-shoot*. Certain tribes of the Orinoco term the pearly drops of dew which cover the beautiful leaves of the heliconia *star-spit*. In the Lithuanian mythology, the nature and signification of falling stars are embodied under nobler and more graceful symbols. The Parcæ, *Werpeja*, weave in heaven for the new-born child its thread of fate, attaching each separate thread to a star. When death approaches the person, the thread is rent, and the star wanes and sinks to the earth.—*Jacob Grimm.*

THEORY AND EXPERIENCE.

In the perpetual vicissitude of theoretical views, says the author of *Giordano Bruno*, "most men see nothing in philosophy but a succession of passing meteors; whilst even the grander forms in which she has revealed herself share the fate of comets,—bodies that do not rank in popular opinion amongst the external and permanent works of nature, but are regarded as mere fugitive apparitions of igneous vapour."

METEORITES FROM THE MOON.

The hypothesis of the selenic origin of meteoric stones depends upon a number of conditions, the accidental coincidence of which could alone convert a possible to an actual fact. The view of the original existence of small planetary masses in space is simpler, and at the same time more analogous with those entertained concerning the formation of other portions of the solar system.

Diogenes Laertius thought aerolites came from the sun; but Pliny derides this theory. The fall of aerolites in bright sunshine, and when the moon's disc was invisible, probably led to the idea of sun-stones. Moreover Anaxagoras regarded the sun as "a molten fiery mass;" and Euripides, in Phaëton, terms the sun "a golden mass," that is to say, a fire-coloured, brightly-shining matter, but not leading to the inference that aerolites are golden sun-stones. The Greek philosophers had four hypotheses as to their origin: telluric, from ascending exhalations; masses of stone raised by hurricanes; a solar origin; and lastly, an origin in the regions of space, as heavenly bodies which had long remained invisible: the last opinion entirely according with that of the present day.

Chladni states that an Italian physicist, Paolo Maria Terzago, on the occasion of the fall of an aerolite at Milan, in 1660, by which a Franciscan monk was killed, was the first who surmised that aerolites were of selenic origin. Without any previous knowledge of this conjecture, Olbers was led, in 1795 (after the celebrated fall at Siena, June 16th, 1794), to investigate the amount of the initial tangential force that would be required to bring to the earth masses projected from the moon. Olbers, Brandes, and Chaldni thought that "the velocity of 16 to 32 miles, with which fire-balls and shooting-stars entered our atmosphere," furnished a refutation to the view of their selenic origin. According to Olbers, it would require to reach the earth, setting aside the resistance of the air, an initial velocity of 8292 feet in the second; according to Laplace, 7862; to Biot, 8282; and to Poisson, 7595. Laplace states that this velocity is only five or six times as great as that of a cannon-ball; but Olbers has shown that "with such an initial velocity as 7500 or 8000 feet in a second, meteoric stones would arrive at the surface of our earth with a velocity of only 35,000 feet." But the measured velocity of meteoric stones averages upwards of 114,000 feet to a second; consequently the original velocity of projection from the moon must be almost 110,000 feet, and therefore 14 times greater than Laplace asserted. It must, however, be recollected, that the opinion then so prevalent, of the existence of active volcanoes in the moon, where air and water are absent, has since been abandoned.

Laplace elsewhere states, that in all probability aerolites "come from the depths of space;" yet he in another passage inclines to the hypothesis of their lunar origin, always, however, assuming that the stones projected from the moon "become satellites of our earth, describing around it more or less eccentric orbits, and thus not reaching its atmosphere until several or even many revolutions have been accomplished."

In Syria there is a popular belief that aerolites chiefly fall on clear moonlight nights. The ancients (Pliny tells us) looked for their fall during lunar eclipses.—*Abridged from Humboldt's Cosmos*, vol. i. (Bohn's edition).

Dr. Laurence Smith, U.S., accepts the "lunar theory," and considers meteorites to be masses thrown off from the moon, the attractive power of which is but one-sixth that of the earth; so that bodies thrown from the surface of the moon experience but one sixth the retarding force they would have when thrown from the earth's surface.

Look again (says Dr. Smith) at the constitution of the meteorite, made up principally of *pure* iron. It came evidently from some place where there is little or no oxygen. Now the moon has no atmosphere, and no water on its surface. There is no oxygen there. Hurled from the moon, these bodies,—these masses of almost pure iron,—would flame in the sun like polished steel, and on reaching our atmosphere would burn in its oxygen until a black oxide cooled it; and this we find to be the case with all meteorites,—the black colour is only an external covering.

Sir Humphry Davy, from facts contained in his researches on flame, in 1817, conceives that the light of meteors depends, not upon the ignition of inflammable gases, but upon that of solid bodies; that such is their velocity of motion, as to excite sufficient heat for their ignition by the compression even of rare air; and that the phenomena of falling stars may be explained by regarding them as small incombustible bodies moving round the earth in very eccentric orbits, and becoming ignited only when they pass with immense rapidity through the upper regions of the atmosphere; whilst those meteors which throw down stony bodies are, similarly circumstanced, combustible masses.

Masses of iron and nickel, having all the appearance of aerolites or meteoric stones, have been discovered in Siberia, at a depth of ten metres below the surface of the earth. From the fact, however, that no meteoric stones are found in the secondary and tertiary formations, it would seem to follow that the phenomena of falling stones did not take place till the earth assumed its present conditions.

VAST SHOWER OF METEORS.

The most magnificent Shower of Meteors that has ever been known was that which fell during the night of November 12th, 1833, commencing at nine o'clock in the evening, and continuing till the morning sun concealed the meteors from view. This shower extended from Canada to the northern boundary of South America, and over a tract of nearly 3000 miles in width.

IMMENSE METEORITE.

Mrs. Somerville mentions a Meteorite which passed within twenty-five miles of our planet, and was estimated to weigh 600,000 tons, and to move with a velocity of twenty miles in a second. Only a small fragment of this immense mass reached the earth. Four instances are recorded of persons being killed by their fall. A block of stone fell at Ægos Potamos, B.C. 465, as large as two millstones; another at Narni, in 921, projected like a rock four feet above the surface of the river, in which it was seen to fall. The Emperor Jehangire had a sword forged from a mass of meteoric iron, which fell in 1620 at Jahlinder in the Punjab. Sixteen instances of the fall of stones in the British Isles are well authenticated to have occurred since 1620, one of them in London. It is very remarkable that no new chemical element has been detected in any of the numerous meteorites which have been analysed.

NO FOSSIL METEORIC STONES.

It is (says Olbers) a remarkable but hitherto unregarded fact, that while shells are found in secondary and tertiary formations, no Fossil Meteoric Stones have as yet been discovered. May we conclude from this circumstance, that previous to the present and last modification of the earth's surface no meteoric stones fell on it, though at the present time it appears probable, from the researches of Schreibers, that 700 fall annually?24

THE END OF OUR SYSTEM.

While all the phenomena in the heavens indicate a law of progressive creation, in which revolving matter is distributed into suns and planets, there are indications in our own system that a period has been assigned for its duration, which, sooner or later, it must reach. The medium which fills universal space, whether it be a luminiferous ether, or arise from the indefinite expansion of planetary atmospheres, must retard the bodies which move in it, even were it 360,000 millions of times more rare than atmospheric air; and, with its time of revolution gradually shortening, the satellite must return to its planet, the planet to its sun, and the sun to its primeval nebula. The fate of our system, thus deduced from mechanical laws, must be the fate of all others. Motion cannot be perpetuated in a resisting medium; and where there exist disturbing forces, there must be primarily derangement, and ultimately ruin. From the great central mass, heat may again be summoned to exhale nebulous matter; chemical forces may again produce motion, and motion may again generate systems; but, as in the recurring catastrophes which have desolated our earth, the great First Cause must preside at the dawn of each cosmical cycle; and, as in the animal races which were successively reproduced, new celestial creations of a nobler form of beauty and of a higher form of permanence may yet appear in the sidereal universe. "Behold, I create new heavens and a new earth, and the former shall not be remembered." "The new heavens and the new earth shall remain before me." "Let us look, then, according to this promise, for the new heavens and the new earth, wherein dwelleth righteousness."—*North-British Review*, No. 3.

BENEFITS OF GLASS TO MAN.

Cuvier eloquently says: "It could not be expected that those Phœnician sailors who saw the sand of the shores of Bætica transformed by fire into a transparent Glass, should have at once foreseen that this new substance would prolong the pleasures of sight to the old; that it would one day assist the astronomer in penetrating the depths of the heavens, and in numbering the stars of the Milky Way; that it would lay open to the naturalist a miniature world, as populous, as rich in wonders as that which alone seemed to have been granted to his senses and his contemplation: in fine, that the most simple and direct use of it would enable the inhabitants of the coast of the Baltic Sea to build palaces more magnificent than those of Tyre and Memphis, and to cultivate, almost under the polar circle, the most delicious fruit of the torrid zone."

THE GALILEAN TELESCOPE.

Galileo appears to be justly entitled to the honour of having invented that form of Telescope which still bears his name; while we must accord to John Lippershey, the spectacle-maker of Middleburg, the honour of having previously invented the astronomical telescope. The interest excited at Venice by Galileo's invention amounted almost to frenzy. On ascending the tower of St. Mark, that he might use one of his telescopes without molestation, Galileo was recognised by a crowd in the street, who took possession of the wondrous tube, and detained the impatient philosopher for several hours, till they had successively witnessed its effects. These instruments were soon manufactured in great numbers; but were purchased merely as philosophical toys, and were carried by travellers into every corner of Europe.

WHAT GALILEO FIRST SAW WITH HIS TELESCOPE.

The moon displayed to him her mountain-ranges and her glens, her continents and her highlands, now lying in darkness, now brilliant with sunshine, and undergoing all those variations of light and shadow which the surface of our own globe presents to the alpine traveller or to the aeronaut. The four satellites of Jupiter illuminating their planet, and suffering eclipses in his shadow, like our own moon; the spots on the sun's disc, proving his rotation round his axis in twenty-five days; the crescent phases of Venus, and the triple form or the imperfectly developed ring of Saturn,—were the other discoveries in the solar system which rewarded the diligence of Galileo. In the starry heavens, too, thousands of new worlds were discovered by his telescope; and the Pleiades alone, which to the unassisted eye exhibit only *seven* stars, displayed to Galileo no fewer than *forty*.—*North-British Review*, No. 3.

The first telescope "the starry Galileo" constructed with a leaden tube a few inches long, with a spectacle-glass, one convex and one concave, at each of its extremities. It magnified three times. Telescopes were made in London in February 1610, a year after Galileo had completed his own (Rigaud, *On Harriot's Papers*, 1833). They were at first called *cylinders*. The telescopes which Galileo constructed, and others of which he made use for observing Jupiter's satellites, the phases of Venus, and the solar spots, possessed the gradually-increasing powers of magnifying four, seven, and thirty-two linear diameters; but they never had a higher power.—Arago, in the *Annuaire* for 1842.

Clock-work is now applied to the equatorial telescope, so as to allow the observer to follow the course of any star, comet, or planet he may wish to observe continuously, without using his hands for the mechanical motion of the instrument.

ANTIQUITY OF TELESCOPES.

Long tubes were certainly employed by Arabian astronomers, and very probably also by the Greeks and Romans; the exactness of their observations being in some degree attributable to their causing the object to be seen through diopters or slits. Abul Hassan speaks very distinctly of tubes, to the extremities of which ocular and object diopters were attached; and instruments so constructed were used in the observatory founded by Hulagu at Meragha. If stars be more easily discovered during twilight by means of tubes, and if a star be sooner revealed to the naked eye through a tube than without it, the reason lies, as Arago has truly observed, in the circumstance that the tube conceals a great portion of the disturbing light diffused in the atmospheric strata between the star and the eye applied to the tube. In like manner, the tube prevents the lateral impression of the faint light which the particles of air receive at night from all the other stars in the firmament. The intensity of the image and the size of the star are apparently augmented.—*Humboldt's Cosmos*, vol. iii. p. 53.

NEWTON'S FIRST REFLECTING TELESCOPE.

The year 1668 may be regarded as the date of the invention of Newton's Reflecting Telescope. Five years previously, James Gregory had described the manner of constructing a reflecting telescope with two concave specula; but Newton perceived the disadvantages to be so great, that, according to his statement, he "found it necessary, before attempting any thing in the practice, to alter the design, and place the eye-glass at the side of the tube rather than at the middle." On this improved principle Newton constructed his telescope, which was examined by Charles II.; it was presented to the Royal Society near the end of 1671, and is carefully preserved by that distinguished body, with the inscription:

" THE FIRST REFLECTING TELESCOPE; INVENTED BY SIR ISAAC NEWTON,

AND MADE WITH HIS OWN HANDS. "

Sir David Brewster describes this telescope as consisting of a concave metallic speculum, the radius of curvature of which was 12-2/3 or 13 inches, so that "it collected the sun's rays at the distance of 6-1/3 inches." The rays reflected by the speculum were received upon a plane metallic speculum inclined 45° to the axis of the tube, so as to reflect them to the side of the tube in which there was an aperture to receive a small tube with a plano-convex eye-glass whose radius was one-twelfth of an inch, by means of which the image formed by the speculum was magnified 38 times. Such was the first reflecting telescope applied to the heavens; but Sir David Brewster describes this instrument as small and ill-made; and fifty years elapsed before telescopes of the Newtonian form became useful in astronomy.

SIR WILLIAM HERSCHEL'S GREAT TELESCOPE AT SLOUGH.

The plan of this Telescope was intimated by Herschel, through Sir Joseph Banks, to George III., who offered to defray the whole expense of it; a noble act of liberality, which has never been imitated by any other British sovereign. Towards the close of 1785, accordingly, Herschel began to construct his reflecting telescope, *forty feet in length*, and having a speculum *fully four feet in diameter*. The thickness of the speculum, which was uniform in every part, was 3½ inches, and its weight nearly 2118 pounds; the metal being composed of 32 copper, and 10•7 of tin: it was the third speculum cast, the two previous attempts having failed. The speculum, when not in use, was preserved from damp by a tin cover, fitted upon a rim of close-grained cloth.

The tube of the telescope was 39 ft. 4 in. long, and its width 4 ft. 10 in.; it was made of iron, and was 3000 lbs. lighter than if it had been made of wood. The observer was seated in a suspended movable seat at the mouth of the tube, and viewed the image of the object with a magnifying lens or eye-piece. The focus of the speculum, or place of the image, was within four inches of the lower side of the mouth of the tube, and came forward into the air, so that there was space for part of the head above the eye, to prevent it from intercepting many of the rays going from the object to the mirror. The eye-piece moved in a tube carried by a slider directed to the centre of the speculum, and fixed on an adjustible foundation at the mouth of the tube. It was completed on the 27th August 1789; and *the very first moment* it was directed to the heavens, a new body was added to the solar system, namely, Saturn and six of its satellites; and in less than a month after, the seventh satellite of Saturn, "an object," says Sir John Herschel, "of a far higher order of difficulty."—*Abridged from the North-British Review*, No. 3.

This magnificent instrument stood on the lawn in the rear of Sir William Herschel's house at Slough; and some of our readers, like ourselves, may remember its extraordinary aspect when seen from the Bath coach-road, and the road to Windsor. The difficulty of managing so large an instrument—requiring as it did two assistants in addition to the observer himself and the person employed to note the time—prevented its being much used. Sir John Herschel, in a letter to Mr. Weld, states the entire cost of its construction, 4000*l.*, was defrayed by George III. In 1839, the woodwork of the telescope being decayed, Sir John Herschel had it cleared away; and piers were erected, on which the tube was placed, *that* being of iron, and so well preserved that, although not more than one-twentieth of an inch thick, when in the horizontal position it contained within all Sir John's family; and next the two reflectors, the polishing apparatus, and portions of the machinery, to the amount of a great many tons. Sir John attributes this great strength and resistance to the internal structure of the tube, very similar to that patented under the name of corrugated iron-roping. Sir John Herschel also thinks that system of triangular arrangement of the woodwork was upon the principle to which "diagonal bracing" owes its strength.

THE EARL OF ROSSE'S GREAT REFLECTING TELESCOPE.

Sir David Brewster has remarked, that "the long interval of half a century seems to be the period of hybernation during which the telescopic mind rests from its labours in order to acquire strength for some great achievement. Fifty years elapsed between the dwarf telescope of Newton and the large instruments of Hadley; other fifty years rolled on before Sir William Herschel constructed his magnificent telescope; and fifty years more passed away before the Earl of Rosse produced that colossal instrument which has already achieved such brilliant discoveries."[25]

In the improvement of the Reflecting Telescope, the first object has always been to increase the magnifying power and light by the construction of as large a mirror as possible; and to this point Lord Rosse's attention was directed as early as 1828, the field of operation being at his lordship's seat, Birr Castle at Parsonstown, about fifty miles west of Dublin. For this high branch of scientific inquiry Lord Rosse was well fitted by a rare combination of "talent to devise, patience to bear disappointment, perseverance, profound mathematical knowledge, mechanical skill, and uninterrupted leisure from other pursuits;"26 all these, however, would not have been sufficient, had not a great command of money been added; the gigantic telescope we are about to describe having cost certainly not less than twelve thousand pounds.

Lord Rosse ground and polished specula fifteen inches, two feet, and three feet in diameter before he commenced the colossal instrument. It is impossible here to detail the admirable contrivances and processes by which he prepared himself for this great work. He first ascertained the most useful combination of metals for specula, both in whiteness, porosity, and hardness, to be copper and tin. Of this compound the reflector was cast in pieces, which were fixed on a bed of zinc and copper,—a species of brass which expanded in the same degree by heat as the pieces of the speculum themselves. They were ground as one body to a true surface, and then polished by machinery moved by a steam-engine. The peculiarities of this mechanism were entirely Lord Rosse's invention, and the result of close calculation and observation: they were chiefly, placing the speculum with the face upward, regulating the temperature by having it immersed in water, usually at 55° Fahr., and regulating the pressure and velocity. This was found to work a perfect spherical figure in large surfaces with a degree of precision unattainable by the hand; the polisher, by working above and upon the face of the speculum, being enabled to examine the operation as it proceeded without removing the speculum, which, when a ton weight, is no easy matter.

The contrivance for doing this is very beautiful. The machine is placed in a room at the bottom of a high tower, in the successive floors of which trap-doors can be opened. A mast is elevated on the top of the tower, so that its summit is about ninety feet *above* the speculum. A dial-plate is attached to the top of the mast, and a small plane speculum and eye-piece, with proper adjustments, are so placed that the combination becomes a Newtonian telescope, and the dial-plate the object. The last and most important part of the process of working the speculum, is to give it a *true parabolic figure*, that is, such a figure that each portion of it should reflect the incident ray to the same focus. Lord Rosse's operations for this purpose consist—1st, of a stroke of the first eccentric, which carries the polisher along *one-third* of the diameter of the speculum; 2d, a transverse stroke twenty-one times slower, and equal to 0•27 of the same diameter, measured on the edge of the tank, or 1•7 beyond the centre of the polisher; 3d, a rotation of the speculum performed in the same time as thirty-seven of the first strokes; and 4th, a rotation of the polisher in the same direction about sixteen times slower. If these rules are attended to, the machine will give the true parabolic figure to the speculum, whether it be *six inches* or *three feet in diameter*. In the three-feet speculum, the figure is so true with the whole aperture, that it is thrown out of focus by a motion of less than the *thirtieth of an inch*, "and even with a single lens of one-eighth of an inch focus, giving a power of 2592, the dots on a watch-dial are still in some degree defined."

Thus was executed the three-feet speculum for the twenty-six-feet telescope placed upon the lawn at Parsonstown, which, in 1840, showed with powers up to 1000 and even 1600; and which resolved nebulæ into stars, and destroyed that symmetry of form in globular nebulæ upon which was founded the hypothesis of the gradual condensation of nebulous matter into suns and planets.27

Scarcely was this instrument out of Lord Rosse's hands, when he resolved to attempt by the same processes to construct another reflector, with a speculum *six feet* in diameter and *fifty feet long*! and this magnificent instrument was completed early in 1845. The focal length of the speculum is fifty-four feet. It weighs four tons, and, with its supports, is seven times as heavy as the four-feet speculum of Sir William Herschel. The speculum is placed in one of the sides of a cubical wooden box, about eight feet wide, and to the opposite end of this box is fastened the tube, which is made of deal staves an inch thick, hooped with iron clamp-rings, like a huge cask. It carries at its upper end, and in the axis of the tube, a small oval speculum, six inches in its lesser diameter.

The tube is about 50 feet long and 8 feet in diameter in the middle, and furnished with diaphragms 6½ feet in aperture. The late Dean of Ely walked through the tube with an umbrella up.

The telescope is established between two lofty castellated piers 60 feet high, and is raised to different altitudes by a strong chain-cable attached to the top of the tube. This cable passes over a pulley on a frame down to a windlass on the ground, which is wrought by two assistants. To the frame are attached chain-guys fastened to the counterweights; and the telescope is balanced by these counterweights suspended by chains, which are fixed to the sides of the tube and pass over large iron pulleys. The immense mass of matter weighs about twelve tons.

On the eastern pier is a strong semicircle of cast-iron, with which the telescope is connected by a racked bar, with friction-rollers attached to the tube by wheelwork, so that by means of a handle near the eye-piece, the observer can move the telescope along the bar on either side of the meridian, to the distance of an hour for an equatorial star.

On the western pier are stairs and galleries. The observing gallery is moved along a railway by means of wheels and a winch; and the mechanism for raising the galleries to various altitudes is very ingenious. Sometimes the galleries, filled with observers, are suspended midway between the two piers, over a chasm sixty feet deep.

An excellent description of this immense Telescope at Birr Castle will be found in Mr. Weld's volume of *Vacation Rambles*.

Sir David Brewster thus eloquently sketches the powers of the telescope at the close of his able description of the instrument, which we have in part quoted from his *Life of Sir Isaac Newton*.

We have, in the mornings, walked again and again, and ever with new delight, along its mystic tube, and at midnight, with its distinguished architect, pondered over the marvellous sights which it dis- closes,—the satellites and belts and rings of Saturn,—the old and new ring, which is advancing with its crest of waters to the body of the planet,—the rocks, and mountains, and valleys, and extinct volcanoes of the moon,—the crescent of Venus, with its mountainous outline,—the systems of double and triple stars,—the nebulæ and starry clusters of every variety of shape,—and those spiral nebular formations which baffle human comprehension, and constitute the greatest achievement in modern discovery.

The Astronomer Royal, Mr. Airy, alludes to the impression made by the enormous light of the telescope,—partly by the modifications produced in the appearance of nebulæ already figured, partly by the great number of stars seen at a distance from the Milky Way, and partly from the prodigious brilliancy of Saturn. The account given by another astronomer of the appearance of Jupiter was that it resembled a coach-lamp in the telescope; and this well expresses the blaze of light which is seen in the instrument.

The Rev. Dr. Scoresby thus records the results of his visits:

The range opened to us by the great telescope at Birr Castle is best, perhaps, apprehended by the now usual measurement—not of distances in miles, or millions of miles, or diameters of the earth's orbit, but—of the progress of light in free space. The determination within, no doubt, a small proportion of error of the parallax of a considerable number of the fixed stars yields, according to Mr. Peters, a space betwixt us and the fixed stars of the smallest magnitude, the sixth, ordinarily visible to the naked eye, of 130 years in the flight of light. This information enables us, on the principles of *sounding the heavens*, suggested by Sir W. Herschel, with the photometrical researches on the stars of Dr. Wollaston and others, to carry the estimation of distances, and that by no means on vague assumption, to the limits of space opened out by the most effective telescopes. And from the guidance thus afforded us as to the comparative power of the six feet speculum in the penetration of space as already elucidated, we might fairly assume the fact, that if any other telescope now in use could follow the sun if removed to the remotest visible position, or till its light would require 10,000 years to reach us, the grand instrument at Parsonstown would follow it so far that from 20,000 to 25,000 years would be spent in the transmission of its light to the earth. But in the cases of clusters of stars, and of nebulæ exhibiting a mere speck of misty luminosity, from the combined light of perhaps hundreds of thousands of suns, the *penetration* into space, compared with the results of ordinary vision, must be enormous; so that it would not be difficult to show the *probability* that a million of years, in flight of light, would be requisite, in regard to the most distant, to trace the enormous interval.

GIGANTIC TELESCOPES PROPOSED.

Hooke is said to have proposed the use of Telescopes having a length of upwards of 10,000 feet (or nearly two miles), in order to see animals in the moon! an extravagant expectation which Auzout considered it necessary to refute. The Capuchin monk Schyrle von Rheita, who was well versed in optics, had already spoken of the speedy practicability of constructing telescopes that should magnify 4000 times, by means of which the lunar mountains might be accurately laid down.

Optical instruments of such enormous focal lengths remind us of the Arabian contrivances of measurement: quadrants with a radius of about 190 feet, upon whose graduated limb the image of the sun was received as in the gnomon, through a small round aperture. Such a quadrant was erected at Samarcand, probably constructed after the model of the older sextants of Alchokandi, which were about sixty feet in height.

LATE INVENTION OF OPTICAL INSTRUMENTS.

A writer in the *North-British Review*, No. 50, considers it strange that a variety of facts which must have presented themselves to the most careless observer should not have led to the earlier construction of Optical Instruments. The ancients, doubtless, must have formed metallic articles with concave surfaces, in which the observer could not fail to see himself magnified; and if the radius of the concavity exceeded twelve inches, twice the focal distance of his eye, he had in his hands an extempore reflecting telescope of the Newtonian form, in which the concave metal was the speculum, and his eye the eye-glass, and which would magnify and bring near him the image of objects nearly behind him. Through the spherical drops of water suspended before his eye, an attentive observer might have seen magnified some minute body placed accidentally in its anterior focus; and in the eyes of fishes and quadrupeds which he used for his food, he might have seen, and might have extracted, the beautiful lenses which they contain, and which he could not fail to regard as the principal agents in the vision of the animals to which they belonged. Curiosity might have prompted him to look through these remarkable lenses or spheres; and had he placed the lens of the smallest minnow, or that of the bird, the sheep, or the ox, in or before a circular aperture, he would have produced a microscope or microscopes of excellent quality and different magnifying powers. No such observations seem, however, to have been made; and even after the invention of glass, and its conversion into globular vessels, through which, when filled with any fluid, objects are magnified, the microscope remained undiscovered.

A TRIAD OF CONTEMPORARY ASTRONOMERS.

It is a remarkable fact in the history of astronomy (says Sir David Brewster), that three of its most distinguished professors were contemporaries. Galileo was the contemporary of Tycho during thirty-seven years, and of Kepler during the fifty-nine years of his life. Galileo was born seven years before Kepler, and survived him nearly the same time. We have not learned that the intellectual triumvirate of the age enjoyed any opportunity for mutual congratulation. What a privilege would it have been to have contrasted the aristocratic dignity of Tycho with the reckless ease of Kepler, and the manly and impetuous mien of the Italian sage!—*Brewster's Life of Newton.*

A PEASANT ASTRONOMER.

At about the same time that Goodricke discovered the variation of the remarkable periodical star Algol, or β Persei, one Palitzch, a farmer of Prolitz, near Dresden,—a peasant by station, an astronomer by nature,—from his familiar acquaintance with the aspect of the heavens, was led to notice, among so many thousand stars, Algol, as distinguished from the rest by its variation, and ascertained its period. The same Palitzch was also the first to re-discover the predicted comet of Halley in 1759, which he saw nearly a month before any of the astronomers, who, armed with their telescopes, were anxiously watching its return. These anecdotes carry us back to the era of the Chaldean shepherds.—*Sir John Herschel's Outlines.*

SHIRBURN-CASTLE OBSERVATORY.

Lord Macclesfield, the eminent mathematician, who was twelve years President of the Royal Society, built at his seat, Shirburn Castle in Oxfordshire, an Observatory, about 1739. It stood 100 yards south from the castle-gate, and consisted of a bed-chamber, a room for the transit, and the third for a mural quadrant. In the possession of the Royal Astronomical Society is a curious print representing two of Lord Macclesfield's servants taking observations in the Shirburn observatory; they are Thomas Phelps, aged 82, who, from being a stable-boy to Lord-Chancellor Macclesfield, rose by his merit and genius to be appointed observer. His companion is John Bartlett, originally a shepherd, in which station he, by books and observation, acquired such a knowledge in computation, and of the heavenly bodies, as to induce Lord Macclesfield to appoint him assistant-observer in his observatory. Phelps was the person who, on December 23d, 1743, discovered the great comet, and made the first observation of it; an account of which is entered in the *Philosophical Transactions*, but not the name of the observer.

LACAILLE'S OBSERVATORY.

Lacaille, who made more observations than all his contemporaries put together, and whose researches will have the highest value as long as astronomy is cultivated, had an observatory at the Collège Mazarin, part of which is now the Palace of the Institute, at Paris.

For a long time it had been without observer or instruments; under Napoleon's reign it was demolished. Lacaille never used to illuminate the wires of his instruments. The inner part of his observatory was painted black; he admitted only the faintest light, to enable him to see his pendulum and his paper: his left eye was devoted to the service of looking to the pendulum, whilst his right eye was kept shut. The latter was only employed to look to the telescope, and during the time of observation never opened but for this purpose. Thus the faintest light made him distinguish the wires, and he very seldom felt the necessity of illuminating them. Part of these blackened walls were visible long after the demolition of the observatory, which took place somewhat about 1811.—*Professor Mohl.*

NICETY REQUIRED IN ASTRONOMICAL CALCULATIONS.

In the *Edinburgh Review*, 1850, we find the following illustrations of the enormous propagation of minute errors:

The rod used in measuring a base-line is commonly about ten feet long; and the astronomer may be said truly to apply that very rod to mete the distance of the stars. An error in placing a fine dot which fixes the length of the rod, amounting to one-five-thousandth of an inch (the thickness of a single silken fibre), will amount to an error of 70 feet in the earth's diameter, of 316 miles in the sun's distance, and to 65,200,000 miles in that of the nearest fixed star. Secondly, as the astronomer in his observatory has nothing further to do with ascertaining lengths or distances, except by calculation, his whole skill and artifice are exhausted in the measurement of angles; for by these alone spaces inaccessible can be compared. Happily, a ray of light is straight: were it not so (in celestial spaces at least), there would be an end of our astronomy. Now an angle of a second (3600 to a degree) is a subtle thing. It has an apparent breadth utterly invisible to the unassisted eye, unless accompanied with so intense a splendour (*e. g.* in the case of a fixed star) as actually to raise by its effect on the nerve of sight a spurious image having a sensible breadth. A silkworm's fibre, such as we have mentioned above, subtends an angle of a second at 3½ feet distance; a cricket-ball, 2½ inches diameter, must be removed, in order to subtend a second, to 43,000 feet, or about 8 miles, where it would be utterly invisible to the sharpest sight aided even by a telescope of some power. Yet it is on the measure of one single second that the ascertainment of a sensible parallax in any fixed star depends; and an error of one-thousandth of that amount (a quantity still unmeasurable by the most perfect of our instruments) would place the star too far or too near by 200,000,000,000 miles; a space which light requires 118 days to travel.

CAN STARS BE SEEN BY DAYLIGHT?

Aristotle maintains that Stars may occasionally be seen in the Daylight, from caverns and cisterns, as through tubes. Pliny alludes to the same circumstance, and mentions that stars have been most distinctly recognised during solar eclipses. Sir John Herschel has heard it stated by a celebrated optician, that his attention was first drawn to astronomy by the regular appearance, at a certain hour, for several successive days, of a considerable star through the shaft of a chimney. The chimney-sweepers who have been questioned upon this subject agree tolerably well in stating that "they have never seen stars by day, but that when observed at night through deep shafts, the sky appeared quite near, and the stars larger." Saussure states that stars have been seen with the naked eye in broad daylight, on the declivity of Mont Blanc, at an elevation of 12,757 feet, as he was assured by several of the alpine guides. The observer must be placed entirely in the shade, and have a thick and massive shade above his head, else the stronger light of the air will disperse the faint image of the stars; these conditions resembling those presented by the cisterns of the ancients, and the chimneys above referred to. Humboldt, however, questions the accuracy of these evidences, adding that in the Cordilleras of Mexico, Quito, and Peru, at elevations of 15,000 or 16,000 feet above the sea-level, he never could distinguish stars by daylight. Yet, under the ethereally pure sky of Cumana, in the plains near the sea-shore, Humboldt has frequently been able, after observing an eclipse of Jupiter's satellites, to find the planet again with the naked eye, and has most distinctly seen it when the sun's disc was from 18° to 20° above the horizon.

LOST HEAT OF THE SUN.

By the nature of our atmosphere, we are protected from the influence of the full flood of solar heat. The absorption of caloric by the air has been calculated at about one-fifth of the whole in passing through a column of 6000 feet, estimated near the earth's surface. And we are enabled, knowing the increasing rarity of the upper regions of our gaseous envelope, in which the absorption is constantly diminishing, to prove that *about one-third of the solar heat is lost* by vertical transmission through the whole extent of our atmosphere.—*J. D. Forbes, F.R.S.*; *Bakerian Lecture*, 1842.

THE LONDON MONUMENT USED AS AN OBSERVATORY.

Soon after the completion of the Monument on Fish Street Hill, by Wren, in 1677, it was used by Hooke and other members of the Royal Society for astronomical purposes, but abandoned on account of the vibrations being too great for the nicety required in their observations. Hence arose *the report that the Monument was unsafe*, which has been revived in our time; "but," says Elmes, "its scientific construction may bid defiance to the attacks of all but earthquakes for centuries to come." This vibration in lofty columns is not uncommon. Captain Smythe, in his *Cycle of Celestial Objects*, tells us, that when taking observations on the summit of Pompey's Pillar, near Alexandria, the mercury was sensibly affected by tremor, although the pillar is a solid.

Geology and Paleontology.

IDENTITY OF ASTRONOMY AND GEOLOGY.

While the Astronomer is studying the form and condition and structure of the planets, in so far as the eye and the telescope can aid him, the Geologist is investigating the form and condition and structure of the planet to which he belongs; and it is from the analogy of the earth's structure, as thus ascertained, that the astronomer is enabled to form any rational conjecture respecting the nature and constitution of the other planetary bodies. Astronomy and Geology, therefore, constitute the same science—the science of material or inorganic nature.

When the astronomer first surveys the *concavity* of the celestial vault, he finds it studded with luminous bodies differing in magnitude and lustre, some moving to the east and others to the west; while by far the greater number seem fixed in space; and it is the business of astronomers to assign to each of them its proper place and sphere, to determine their true distance from the earth, and to arrange them in systems throughout the regions of sidereal space.

In like manner, when the geologist surveys the *convexity* of his own globe, he finds its solid covering composed of rocks and beds of all shapes and kinds, lying at every possible angle, occupying every possible position, and all of them, generally speaking, at the same distance from the earth's centre. Every where we see what was deep brought into visible relation with what was superficial—what is old with what is new—what preceded life with what followed it.

Thus displayed on the surface of his globe, it becomes the business of the geologist to ascertain how these rocks came into their present places, to determine their different ages, and to fix the positions which they originally occupied, and consequently their different distances from the centre or the circumference of the earth. Raised from their original bed, the geologist must study the internal forces by which they were upheaved, and the agencies by which they were indurated; and when he finds that strata of every kind, from the primitive granite to the recent tertiary marine mud, have been thus brought within his reach, and prepared for his analysis, he reads their respective ages in the organic remains which they entomb; he studies the manner in which they have perished, and he counts the cycles of time and of life which they disclose.—*Abridged from the North-British Review*, No. 9.

THE GEOLOGY OF ENGLAND

is more interesting than that of other countries, because our island is in a great measure an epitome of the globe; and the observer who is familiar with our strata, and the fossil remains which they include, has not only prepared himself for similar inquiries in other countries, but is already, as it were, by anticipation, acquainted with what he is to find there.—*Transactions of the Geological Society.*

PROBABLE ORIGIN OF THE ENGLISH CHANNEL.

The proposed construction of a submarine tunnel across the Straits of Dover has led M. Boué, For. Mem. Geol. Soc., to point out the probability that the English Channel has not been excavated by water-action only; but owes its origin to one of the lines of disturbance which have fissured this portion of the earth's crust: and taking this view of the case, the fissure probably still exists, being merely filled with comparatively loose material, so as to prove a serious obstacle to any attempt made to drive through it a submarine tunnel.—*Proceedings of the Geological Society.*

HOW BOULDERS ARE TRANSPORTED TO GREAT HEIGHTS.

Sir Roderick Murchison has shown that in Russia, when the Dwina is at its maximum height, and penetrates into the chinks of its limestone banks, when frozen and expanded it causes disruptions of the rock, the entanglement of stony fragments in the ice. In remarkable spring floods, the stream so expands that in bursting it throws up its icy fragments to 15 or 20 feet above the stream; and the waters subsiding, these lateral ice-heaps melt away, and leave upon the bank the rifled and angular blocks as evidence of the highest ice-mark. In Lapland, M. Böhtlingk assures us that he has found *large granitic boulders weighing several tons actually entangled and suspended, like birds'-nests, in the branches of pine-trees, at heights of 30 or 40 feet above the summer level of the stream!*28

WHY SEA-SHELLS ARE FOUND AT GREAT HEIGHTS.

The action of subterranean forces in breaking through and elevating strata of sedimentary rocks,—of which the coast of Chili, in consequence of a great earthquake, furnishes an example,—leads to the assumption that the pelagic shells found by MM. Bonpland and Humboldt on the ridge of the Andes, at an elevation of more than 15,000 English feet, may have been conveyed to so extraordinary a position, not by a rising of the ocean, but by the agency of volcanic forces capable of elevating into ridges the softened crust of the earth.

SAND OF THE SEA AND DESERT.

That sand is an assemblage of small stones may be seen with the eye unarmed with art; yet how few are equally aware of the synonymous nature of the sand of the sea and of the land! Quartz, in the form of sand, covers almost entirely the bottom of the sea. It is spread over the banks of rivers, and forms vast plains, even at a very considerable elevation above the level of the sea, as the desert of Sahara in Africa, of Kobi in Asia, and many others. This quartz is produced, at least in part, from the disintegration of the primitive granite rocks. The currents of water carry it along, and when it is in very small, light, and rounded grains, even the wind transports it from one place to another. The hills are thus made to move like waves, and a deluge of sand frequently inundates the neighbouring countries:

"So where o'er wide Numidian wastes extend,Sudden the impetuous hurricanes descend."—*Addison's Cato.*

To illustrate the trite axiom, that nothing is lost, let us glance at the most important use of sand:

"Quartz in the form of sand," observes Maltebrun, "furnishes, by fusion, one of the most useful substances we have, namely glass, which, being less hard than the crystals of quartz, can be made equally transparent, and is equally serviceable to our wants and to our pleasures. There it shines in walls of crystal in the palaces of the great, reflecting the charms of a hundred assembled beauties; there, in the hand of the philosopher, it discovers to us the worlds that revolve above us in the immensity of space, and the no less astonishing wonders that we tread beneath our feet."

PEBBLES.

The various heights and situations at which Pebbles are found have led to many erroneous conclusions as to the period of changes of the earth's surface. All the banks of rivers and lakes, and the shores of the sea, are covered with pebbles, rounded by the waves which have rolled them against each other, and which frequently seem to have brought them from a distance. There are also similar masses of pebbles found at very great elevations, to which the sea appears never to have been able to reach. We find them in the Alps at Valorsina, more than 6000 feet above the level of the sea; and on the mountain of Bon Homme, which is more than 1000 feet higher. There are some places little elevated above the level of the sea, which, like the famous plain of Crau, in Provence, are entirely paved with pebbles; while in Norway, near Quedlia, some mountains of considerable magnitude seem to be completely formed of them, and in such a manner that the largest pebbles occupy the summit, and their thickness and size diminish as you approach the base. We may include in the number of these confused and irregular heaps most of the depositions of matter brought by the river or sea, and left on the banks, and perhaps even those immense beds of sand which cover the centre of Asia and Africa. It is this circumstance which renders so uncertain the distinction, which it is nevertheless necessary to establish, between alluvial masses created before the commencement of history, and those which we see still forming under our own eyes.

A charming monograph, entitled "Thoughts on a Pebble," full of playful sentiment and graceful fancy, has been written by the amiable Dr. Mantell, the geologist.

ELEVATION OF MOUNTAIN-CHAINS.

Professor Ansted, in his *Ancient World*, thus characterises this phenomenon:

These movements, described in a few words, were doubtless going on for many thousands and tens of thousands of revolutions of our planet. They were accompanied also by vast but slow changes of other kinds. The expansive force employed in lifting up, by mighty movements, the northern portion of the continent of Asia, found partial vent; and from partial subaqueous fissures there were poured out the tabular masses of basalt occurring in Central India; while an extensive area of depression in the Indian Ocean, marked by the coral islands of the Laccadives, the Maldives, the great Chagos bank, and some others, were in the course of depression by a counteracting movement.

Hitherto the processes of denudation and of elevation have been so far balanced as to preserve a pretty steady proportion of sea and dry land during geological ages; but if the internal temperature should be so far reduced as to be no longer capable of generating forces of expansion sufficient for this elevatory action, while the denuding forces should continue to act with unabated energy, the inevitable result would be, that every mountain-top would be in time brought low. No earthly barrier could declare to the ocean that there its proud waves should be stayed. Nothing would stop its ravages till all dry land should be laid prostrate, to form the bed over which it would continue to roll an uninterrupted sea.

THE CHALK FORMATION.

Mr. Horner, F.R.S., among other things in his researches in the Delta, considers it extremely probable that every particle of Chalk in the world has at some period been circulating in the system of a living animal.

WEAR OF BUILDING-STONES.

Professor Henry, in an account of testing the marbles used in building the Capitol at Washington, states that every flash of lightning produces an appreciable amount of nitric acid, which, diffused in rain-water, acts on the carbonate of lime; and from specimens subjected to actual freezing, it was found that in ten thousand years one inch would be worn from the blocks by the action of frost.

In 1839, a report of the examination of Sandstones, Limestones, and Oolites of Britain was made to the Government, with a view to the selection of the best material for building the new Houses of Parliament. For this purpose, 103 quarries were described, 96 buildings in England referred to, many chemical analyses of the stones were given, and a great number of experiments related, showing, among other points, the cohesive power of each stone, and the amount of disintegration apparent, when subjected to Brard's process. The magnesian limestone, or dolomite of Bolsover Moor, was recommended, and finally adopted for the Houses; but the selection does not appear to have been so successful as might have been expected from the skill and labour of the investigation. It may be interesting to add, that the publication of the above Report (for which see *Year-Book of Facts*, 1840, pp. 78–80) occasioned Mr. John Mallcott to remark in the *Times* journal, "that all stone made use of in the immediate neighbourhood of its own quarries is more likely to endure that atmosphere than if it be removed therefrom, though only thirty or forty miles:" and the lapse of comparatively few years has proved the soundness of this observation.29

PHENOMENA OF GLACIERS ILLUSTRATED.

Professor Tyndall, being desirous of investigating some of the phenomena presented by the large masses of mountain-ice,—those frozen rivers called Glaciers,—devised the plan of sending a destructive agent into the midst of a mass of ice, so as to break down its structure in the interior, in order to see if this method would reveal any thing of its internal constitution. Taking advantage of the bright weather of 1857, he concentrated a beam of sunlight by a condensing lens, so as to form the focus of the sun's rays in the midst of a mass of ice. A portion of the ice was melted, but the surrounding parts shone out as brilliant stars, produced by the reflection of the faces of the crystalline structure. On examining these brilliant portions with a lens, Professor Tyndall discovered that the structure of the ice had been broken down in symmetrical forms of great beauty, presenting minute stars, surrounded by six petals, forming a beautiful flower, the plane being always parallel to the plane of congelation of the ice. He then prepared a piece of ice, by making both its surfaces smooth and parallel to each other. He concentrated in the centre of the ice the rays of heat from the electric light; and then, placing the piece of ice in the electric microscope, the disc revealed these beautiful ice-flowers.

A mass of ice was crushed into fragments; the small fragments were then placed in a cup of wood; a hollow wooden die, somewhat smaller than the cup, was then pressed into the cup of ice-fragments by the pressure of a hydraulic press, and the ice-fragments were immediately united into a compact cup of nearly transparent ice. This pressure of fragments of ice into a solid mass explains the formation of the glaciers and their origin. They are composed of particles of ice or snow; as they descend the sides of the mountain, the pressure of the snow becomes sufficiently great to compress the mass into solid ice, until it becomes so great as to form the beautiful blue ice of the glaciers. This compression, however, will not form the solid mass unless the temperature of the ice be near that of freezing water. To prove this, the lecturer cooled a mass of ice, by wrapping it in a piece of tinfoil and exposing it for some time to a bath of the ethereal solution of solidified carbonic-acid gas, the coldest freezing mixture known. This cooled mass of ice was crushed to fragments, and submitted to the same pressure which the other fragments had been exposed to without cohering in the slightest degree.—*Lecture at the Royal Institution*, 1858.

ANTIQUITY OF GLACIERS.

The importance of glacier agency in the past as well as the present condition of the earth, is undoubtedly very great. One of our most accomplished and ingenious geologists has, indeed, carried back the existence of Glaciers to an epoch of dim antiquity, even in the reckoning of that science whose chronology is counted in millions of years. Professor Ramsay has shown ground for believing that in the fragments of rock that go to make up the conglomerates of the Permian strata, intermediate between the Old and the New Red Sandstone, there is still preserved a record of the action of ice, either in glaciers or floating icebergs, before those strata were consolidated.—*Saturday Review*, No. 142.

FLOW OF THE MER DE GLACE.

Michel Devouasson of Chamouni fell into a crevasse on the Glacier of Talefre, a feeder of the Mer de Glace, on the 29th of July 1836, and after a severe struggle extricated himself, leaving his knapsack below. The identical knapsack reappeared in July 1846, at a spot on the surface of the glacier *four thousand three hundred* feet from the place where it was lost, as ascertained by Professor Forbes, who himself collected the fragments; thus indicating the rate of flow of the icy river in the intervening ten years.—*Quarterly Review*, No. 202.

THE ALLUVIAL LAND OF EGYPT: ANCIENT POTTERY.

Mr. L. Horner, in his recent researches near Cairo, with the view of throwing light upon the geological history of the alluvial land of Egypt, obtained from the lowest part of the boring of the sediment at the colossal statue of Rameses, at a depth of thirty-nine feet, this curious relic of the ancient world; the boring instrument bringing up a fragment of pottery about an inch square and a quarter of an inch in thickness— the two surfaces being of a brick-red colour, the interior dark gray. According to Mr. Horner's deductions, this fragment, having been found at a depth of 39 feet (if there be no fallacy in his reasoning), must be held to be a record of the existence of man 13,375 years before A.D. 1858, reckoning by the calculated rate of increase of three inches and a half of alluvium in a century—11,517 years before the Christian era, and 7625 before the beginning assigned by Lepsius to the reign of Menos, the founder of Memphis. Moreover it proves in his opinion, that man had already reached a state of civilisation, so far at least as to be able to fashion clay into vessels, and to know how to harden it by the action of strong heat. This calculation is supported by the Chevalier Bunsen, who is of opinion that the first epochs of the history of the human race demand at the least a period of 20,000 years before our era as a fair starting-point in the earth's history.—*Proceedings of Royal Soc.*, 1858.

Upon this theory, a Correspondent, "An Old Indigo-Planter," writes to the *Athenæum*, No. 1509, the following suggestive note: "Having lived many years on the banks of the Ganges, I have seen the stream encroach on a village, undermining the bank where it stood, and deposit, as a natural result, bricks, pottery, &c. in the bottom of the stream. On one occasion, I am certain that the depth of the stream where the bank was breaking was above 40 feet; yet in three years the current of the river drifted so much, that a fresh deposit of soil took place over the *débris* of the village, and the earth was raised to a level with the old bank. Now had our traveller then obtained a bit of pottery from where it had lain for only three years, could he reasonably draw the inference that it had been made 13,000 years before?"

SUCCESSIVE CHANGES OF THE TEMPLE OF SERAPIS.

The Temple of Serapis at Puzzuoli, near Naples, is perhaps, of all the structures raised by the hands of man, the one which affords most instruction to a geologist. It has not only undergone a wonderful succession of changes in past time, but is still undergoing changes of condition. This edifice was exhumed in 1750 from the eastern shore of the Bay of Baiæ, consisting partly of strata containing marine shells with fragments of pottery and sculpture, and partly of volcanic matter of sub-aerial origin. Various theories were proposed in the last century to explain the perforations and attached animals observed on the middle zone of the three erect marble columns until recently standing; Goethe, among the rest, suggesting that a lagoon had once existed in the vestibule of the temple, filled during a temporary incursion of the sea with salt water, and that marine mollusca and annelids flourished for years in this lagoon at twelve feet or more above the sea-level.

This hypothesis was advanced at a time when almost any amount of fluctuation in the level of the sea was thought more probable than the slightest alteration in the level of the solid land. In 1807 the architect Niccolini observed that the pavement of the temple was dry, except when a violent south wind was blowing; whereas, on revisiting the temple fifteen years later, he found the pavement covered by salt water twice every day at high tide. From measurements made from 1822 to 1838, and thence to 1845, he inferred that the sea was gaining annually upon the floor of the temple at the rate of about one-third of an inch during the first period, and about three-fourths of an inch during the second. Mr. Smith of Jordan Hill, from his visits in 1819 and 1845, found an average rise of about an inch annually, which was in accordance with visits made by Mr. Babbage in 1828, and Professor James Forbes in 1826 and 1843. In 1852 Signor Scaecchi, at the request of Sir Charles Lyell, compared the depth of water on the pavement with its level taken by him in 1839, and found that it had gained only 4½ inches in thirteen years, and was not so deep as when MM. Niccolini and Smith measured it in 1845; from which he inferred that after 1845 the downward movement of the land had ceased, and before 1852 had been converted into an upward movement.

Arago and others maintained that the surface on which the temple stands has been depressed, has *remained under the sea, and has again been elevated*. Russager, however, contends that there is nothing in the vicinity of the temple, or in the temple itself, to justify this bold hypothesis. Every thing leads to the belief that the temple has remained unchanged in the position in which it was originally built; but that the sea rose, surrounded it to a height of at least twelve feet, and again retired; but the elevated position of the sea continued sufficiently long to admit of the animals boring the pillars. This view can even be proved historically; for Niccolini, in a memoir published in 1840, gives the heights of the level of the sea in the Bay of Naples for a period of 1900 years, and has with much acuteness proved his assertions historically. The correctness of Russager's opinion, he states, can be demonstrated and reduced to figures by means of the dates collected by Niccolini.—See *Jameson's Journal*, No. 58.

At the present time the floor is always covered with sea-water. On the whole, there is little doubt that the ground has sunk upwards of two feet during the last half-century.

This gradual subsidence confirms in a remarkable manner Mr. Babbage's conclusions—drawn from the calcareous incrustations formed by the hot springs on the walls of the building and from the ancient lines of the water-level at the base of the three columns—that the original subsidence was not sudden, but slow and by successive movements.

Sir Charles Lyell (who, in his *Principles of Geology*, has given a detailed account of the several upfillings of the temple) considers that when the mosaic pavement was re-constructed, the floor of the building must have stood about twelve feet above the level of 1838 (or about 11½ feet above the level of the sea), and that it had sunk about nineteen feet below that level before it was elevated by the eruption of Monte Nuovo.

We regret to add, that the columns of the temple are no longer in the position in which they served so many years as a species of self-registering hydrometer: the materials have been newly arranged, and thus has been torn as it were from history a page which can never be replaced.

THE GROTTO DEL CANE.

This "Dog Grotto" has been so much cited for its stratum of carbonic-acid gas covering the floor, that all geological travellers who visit Naples feel an interest in seeing the wonder.

This cavern was known to Pliny. It is continually exhaling from its sides and floor volumes of steam mixed with carbonic-acid gas; but the latter, from its greater specific gravity, accumulates at the bottom, and flows over the step of the door. The upper part of the cave, therefore, is free from the gas, while the floor is completely covered by it. Addison, on his visit, made some interesting experiments. He found that a pistol could not be fired at the bottom; and that on laying a train of gunpowder and igniting it on the outside of the cavern, the carbonic-acid gas "could not intercept the train of fire when it once began flashing, nor hinder it from running to the very end." He found that a viper was nine minutes in dying on the first trial, and ten minutes on the second; this increased vitality being, in his opinion, attributable to the stock of air which it had inhaled after the first trial. Dr. Daubeny found that phosphorus would continue lighted at about two feet above the bottom; that a sulphur-match went out in a few minutes above it, and a wax-taper at a still higher level. The keeper of the cavern has a dog, upon which he shows the effects of the gas, which, however, are quite as well, if not better, seen in a torch, a lighted candle, or a pistol.

"Unfortunately," says Professor Silliman, "like some other grottoes, the enchantment of the 'Dog Grotto' disappears on a near view." It is a little hole dug artificially in the side of a hill facing Lake Agnano: it is scarcely high enough for a person to stand upright in, and the aperture is closed by a door. Into this narrow cell a poor little dog is very unwillingly dragged and placed in a depression of the floor, where he is soon narcotised by the carbonic acid. The earth is warm to the hand, and the gas given out is very constant.

THE WATERS OF THE GLOBE GRADUALLY DECREASING.

This was maintained by M. Bory Saint Vincent, because the vast deserts of sand, mixed up with the salt and remains of marine animals, of which the surface of the globe is partly composed, were formerly inland seas, which have insensibly become dry. The Caspian, the Dead Sea, the Lake Baikal, &c. will become dry in their turn also, when their beds will be sandy deserts. The inland seas, whether they have only one outlet, as the Mediterranean, the Red Sea, the Baltic, &c., or whether they have several, as the Gulf of Mexico, the seas of O'Kotsk, of Japan, China, &c., will at some future time cease to communicate with the great basins of the ocean; they will become inland seas, true Caspians, and in due time will become likewise dry. On all sides the waters of rivers are seen to carry forward in their course the soil of the continent. Alluvial lands, deltas, banks of sand, form themselves near the coasts, and in the directions of the currents; madreporic animals lay the foundations of new lands; and while the straits become closed, while the depths of the sea fill up, the level of the sea, which it would seem natural should become higher, is sensibly lower. There is, therefore, an actual diminution of liquid matter.

THE SALT LAKE OF UTAH.

Lieutenant Gunnison, who has surveyed the great basin of the Salt Lake, states the water to be about one-third salt, which it yields on boiling. Its density is considerably greater than that of the Red Sea. One can hardly get the whole body below the surface: in a sitting position the head and shoulders will remain above the water, such is the strength of the brine; and on coming to the shore the body is covered with an incrustation of salt in fine crystals. During summer the lake throws on shore abundance of salt, while in winter it throws up Glauber salt plentifully. "The reason of this," says Lieutenant Gunnison, "is left for the scientific to judge, and also what becomes of the enormous amount of fresh water poured into it by three or four large rivers,—Jordan, Bear, and Weber,—as there is no visible effect."

FORCE OF RUNNING WATER.

It has been proved by experiment that the rapidity at the bottom of a stream is every where less than in any other part of it, and is greatest at the surface. Also, that in the middle of the stream the particles at the top move swifter than those at the sides. This slowness of the lowest and side currents is produced by friction; and when the rapidity is sufficiently great, the soil composing the sides and bottom gives way. If the water flows at the rate of three inches per second, it will tear up fine clay; six inches per second, fine sand; twelve inches per second, fine gravel; and three feet per second, stones the size of an egg.—*Sir Charles Lyell.*

THE ARTESIAN WELL OF GRENELLE AT PARIS.

M. Peligot has ascertained that the Water of the Artesian Well of Grenelle contains not the least trace of air. Subterranean waters ought therefore to be *aerated* before being used as aliment. Accordingly, at Grenelle, has been constructed a tower, from the top of which the water descends in innumerable threads, so as to present as much surface as possible to the air.

The boring of this Well by the Messrs. Mulot occupied seven years, one month, twenty-six days, to the depth of 1794½ English feet, or 194½ feet below the depth at which M. Elie de Beaumont foretold that water would be found. The sound, or borer, weighed 20,000 lb., and was treble the height of that of the dome of the Hôpital des Invalides at Paris. In May 1837, when the bore had reached 1246 feet 8 inches, the great chisel and 262 feet of rods fell to the bottom; and although these weighed five tons, M. Mulot tapped a screw on the head of the rods, and thus, connecting another length to them, after fifteen months' labour, drew up the chisel. On another occasion, this chisel having been raised with great force, sank at one stroke 85 feet 3 inches into the chalk!

The depth of the Grenelle Well is nearly four times the height of Strasburg Cathedral; more than six times the height of the Hôpital des Invalides at Paris; more than four times the height of St. Peter's at Rome; nearly four times and a half the height of St. Paul's, and nine times the height of the Monument, London. Lastly, suppose all the above edifices to be piled one upon each other, from the base-line of the Well of Grenelle, and they would but reach within 11½ feet of its surface.

MM. Elie de Beaumont and Arago never for a moment doubted the final success of the work; their confidence being based on analogy, and on a complete acquaintance with the geological structure of the Paris basin, which is identical with that of the London basin beneath the London clay.

In the duchy of Luxembourg is a well the depth of which surpasses all others of the kind. It is upwards of 1000 feet more than that of Grenelle near Paris.

HOW THE GULF-STREAM REGULATES THE TEMPERATURE OF LONDON.

Great Britain is almost exactly under the same latitude as Labrador, a region of ice and snow. Apparently, the chief cause of the remarkable difference between the two climates arises from the action of the great oceanic Gulf-Stream, whereby this country is kept constantly encircled with waters warmed by a West-Indian sun.

Were it not for this unceasing current from tropical seas, London, instead of its present moderate average winter temperature of 6° above the freezing-point, might for many months annually be ice-bound by a settled cold of 10° to 30° below that point, and have its pleasant summer months replaced by a season so short as not to allow corn to ripen, or only an alpine vegetation to flourish.

Nor are we without evidence afforded by animal life of a greater cold having prevailed in this country at a late geological period. One case in particular occurs within eighty miles of London, at the village of Chillesford, near Woodbridge, where, in a bed of clayey sand of an age but little (geologically speaking) anterior to the London gravel, Mr. Prestwich has found a group of fossil shells in greater part identical with species now living in the seas of Greenland and of similar latitudes, and which must evidently, from their perfect condition and natural position, have existed in the place where they are now met with.—*Lectures on the Geology of Clapham, &c. by Joseph Prestwich, A.R.S., F.G.S.*

SOLVENT ACTION OF COMMON SALT AT HIGH TEMPERATURES.

Forchhammer, after a long series of experiments, has come to the conclusion that Common Salt at high temperatures, such as prevailed at earlier periods of the earth's history, acted as a general solvent, similarly to water at common temperatures. The amount of common salt in the earth would suffice to cover its whole surface with a crust ten feet in thickness.

FREEZING CAVERN IN RUSSIA.

This famous Cavern, at Ithetz Kaya-Zastchita, in the Steppes of the Kirghis, is employed by the inhabitants as a cellar. It has the very remarkable property of being so intensely cold during the hottest summers as to be then filled with ice, which disappearing with cold weather, is entirely gone in winter, when all the country is clad in snow. The roof is hung with ever-dripping solid icicles, and the floor may be called a stalagmite of ice and frozen earth. "If," says Sir R. Murchison, "as we were assured, *the cold is greatest when the external air is hottest and driest*, that the fall of rain and a moist atmosphere produce some diminution of the cold in the cave, and that upon the setting-in of winter the ice disappears entirely,—then indeed the problem is very curious." The peasants assert that in winter they could sleep in the cave without their sheepskins.

INTERIOR TEMPERATURE OF THE EARTH: CENTRAL HEAT.

By the observed temperature of mines, and that at the bottom of artesian wells, it has been established that the rate at which such temperature increases as we descend varies considerably in different localities, where the depths are comparatively small; but where the depths are great, we find a much nearer approximation to a common rate of increase, which, as determined by the best observation in the deepest mines, shafts, and artesian wells in Western Europe, is very nearly 1° F. *for an increase in depth of fifty feet.—W. Hopkins, M.A., F.R.S.*

Humboldt states that, according to tolerably coincident experiments in artesian wells, it has been shown that the heat increases on an average about 1° for every 54•5 feet. If this increase can be reduced to arithmetical relations, it will follow that a stratum of granite would be in a state of fusion at a depth of nearly twenty-one geographical miles, or between four and five times the elevation of the highest summit of the Himalaya.

The following is the opinion of Professor Silliman:

That the whole interior portion of the earth, or at least a great part of it, is an ocean of melted rock, agitated by violent winds, though I dare not affirm it, is still rendered highly probable by the phenomena of volcanoes. The facts connected with their eruption have been ascertained and placed beyond a doubt. How, then, are they to be accounted for? The theory prevalent some years since, that they are caused by the combustion of immense coal-beds, is puerile and now entirely abandoned. All the coal in the world could not afford fuel enough for one of the tremendous eruptions of Vesuvius.

This observed increase of temperature in descending beneath the earth's surface suggested the notion of a central incandescent nucleus still remaining in a state of fluidity from its elevated temperature. Hence the theory that the whole mass of the earth was formerly a molten fluid mass, the exterior portion of which, to some unknown depth, has assumed its present solidity by the radiation of heat into surrounding space, and its consequent refrigeration.

The mathematical solution of this problem of Central Heat, assuming such heat to exist, tells us that though the central portion of the earth may consist of a mass of molten matter, the temperature of its surface is not thereby increased by more than the small fraction of a degree. Poisson has calculated that it would require *a thousand millions of centuries* to reduce this fraction to a degree by half its present amount, supposing always the external conditions to remain unaltered. In such cases, the superficial temperature of the earth may, in fact, be considered to have approximated so near to its ultimate limit that it can be subject to no further sensible change.

DISAPPEARANCE OF VOLCANIC ISLANDS.

Many of the Volcanic Islands thrown up above the sea-level soon disappear, because the lavas and conglomerates of which they are formed spread over flatter surfaces, through the weight of the incumbent fluid; and the constant levelling process goes on below the sea by the action of tides and currents. Such islands as have effectually resisted this action are found to possess a solid framework of lava, supporting or defending the loose fragmentary materials.

Among the most celebrated of these phenomena in our times may be mentioned the Isle of Sabrina, which rose off the coast of St. Michael's in 1811, attained a circumference of one mile and a height of 300 feet, and disappeared in less than eight months; in the following year there were eighty fathoms of water in its place. In July 1831 appeared Graham's Island off the coast of Sicily, which attained a mile in circumference and 150 or 160 feet in height; its formation much resembled that of Sabrina.

The line of ancient subterranean fire which we trace on the Mediterranean coasts has had a strange attestation in Graham's Island, which is also described as a volcano suddenly bursting forth in the mid sea between Sicily and Africa; burning for several weeks, and throwing up an isle, or crater-cone of scoriæ and ashes, which had scarcely been named before it was again lost by subsidence beneath the sea, leaving only a shoal-bank to attest this strange submarine breach in the earth's crust, which thus mingled fire and water in one common action.

Floating islands are not very rare: in 1827, one was seen twenty leagues to the east of the Azores; it was three leagues in width, and covered with volcanic products, sugar-canes, straw, and pieces of wood.

PERPETUAL FIRE.

Not far from the Deliktash, on the side of a mountain in Lycia, is the Perpetual Fire described some forty years since by Captain Beaufort. It was found by Lieutenant Spratt and Professor Forbes, thirty years later, as brilliant as ever, and somewhat increased; for besides the large flame in the corner of the ruins described by Beaufort, there were small jets issuing from crevices in the side of the crater-like cavity five or six feet deep. At the bottom was a shallow pool of sulphureous and turbid water, regarded by the Turks as a sovereign remedy for all skin complaints. The soot deposited from the flames was held to be efficacious for sore eyelids, and valued as a dye for the eyebrows. This phenomenon is described by Pliny as the flame of the Lycian Chimera.

ARTESIAN FIRE-SPRINGS IN CHINA.

According to the statement of the missionary Imbert, the Fire-Springs, "Ho-tsing" of the Chinese, which are sunk to obtain a carburetted-hydrogen gas for salt-boiling, far exceed our artesian springs in depth. These springs are very commonly more than 2000 feet deep; and a spring of continued flow was found to be 3197 feet deep. This natural gas has been used in the Chinese province Tse-tschuan for several thousand years; and "portable gas" (in bamboo-canes) has for ages been used in the city of Khiung-tscheu. More recently, in the village of Fredonia, in the United States, such gas has been used both for cooking and for illumination.

VOLCANIC ACTION THE GREAT AGENT OF GEOLOGICAL CHANGE.

Mr. James Nasmyth observes, that "the floods of molten lava which volcanoes eject are nothing less than remaining portions of what was once the condition of the entire globe when in the igneous state of its early physical history,—no one knows how many years ago!

"When we behold the glow and feel the heat of molten lava, how vastly does it add to the interest of the sight when we consider that the heat we feel and the light we see are the residue of the once universal condition of our entire globe, on whose *cooled surface* we *now* live and have our being! But so it is; for if there be one great fact which geological research has established beyond all doubt, it is that we reside on the cooled surface of what was once a molten globe, and that all the phenomena which geology has brought to light can be most satisfactorily traced to the successive changes incidental to its gradual cooling and contraction.

"That the influx of the sea into the yet hot and molten interior of the globe may occasionally occur, and enhance and vary the violence of the phenomenon of volcanic action, there can be little doubt; but the action of water in such cases is only *secondary*. But for the pre-existing high temperature of the interior of the earth, the influx of water would produce no such discharges of molten lava as generally characterise volcanic eruptions. Molten lava is therefore a true vestige of the Natural History of the Creation."

THE SNOW-CAPPED VOLCANO.

It is but rarely that the elastic forces at work within the interior of our globe have succeeded in breaking through the spiral domes which, resplendent in the brightness of eternal snow, crown the summits of the Cordilleras; and even where these subterranean forces have opened a permanent communication with the atmosphere, through circular craters or long fissures, they rarely send forth currents of lava, but merely eject ignited scoriæ, steam, sulphuretted hydrogen gas, and jets of carbonic acid.—*Humboldt's Cosmos*, vol. i.

TRAVELS OF VOLCANIC DUST.

On the 2d of September 1845, a quantity of Volcanic Dust fell in the Orkney Islands, which was supposed to have originated in an eruption of Hecla, in Iceland. It was subsequently ascertained that an eruption of that volcano took place on the morning of the above day (September 2), so as to leave no doubt of the accuracy of the conclusion. The dust had thus travelled about 600 miles!

GREAT ERUPTIONS OF VESUVIUS.

In the great eruption of Vesuvius, in August 1779, which Sir William Hamilton witnessed from his villa at Pausilippo in the bay of Naples, the volcano sent up white sulphureous smoke resembling bales of cotton, exceeding the height and size of the mountain itself at least four times; and in the midst of this vast pile of smoke, stones, scoriæ, and ashes were thrown up not less than 2000 feet. Next day a fountain of fire shot up with such height and brilliancy that the smallest objects could be clearly distinguished at any place within six miles or more of Vesuvius. But on the following day a more stupendous column of fire rose three times the height of Vesuvius (3700 feet), or more than two miles high. Among the huge fragments of lava thrown out during this eruption was a block 108 feet in circumference and 17 feet high, another block 66 feet in circumference and 19 feet high, and another 16 feet high and 92 feet in circumference, besides thousands of smaller fragments. Sir William Hamilton suggests that from a scene of the above kind the ancient poets took their ideas of the giants waging war with Jupiter.

The eruption of June 1794, which destroyed the greater part of the town of Torre del Greco, was, however, the most violent that has been recorded after the two great eruptions of 79 and 1631.

EARTH-WAVES.

The waves of an earthquake have been represented in their progress, and their propagation, through rocks of different density and elasticity; and the causes of the rapidity of propagation, and its diminution by the refraction, reflection, and interference of the oscillations have been mathematically investigated. Air, water, and earth waves follow the same laws which are recognised by the theory of motion, at all events in space; but the earth-waves are accompanied in their destructive action by discharges of elastic vapours, and of gases, and mixtures of pyroxene crystals, carbon, and infusorial animalcules with silicious shields. The more terrific effects are, however, when the earth-waves are accompanied by cleavage; and, as in the earthquake of Riobamba, when fissures alternately opened and closed again, so that men saved themselves by extending both arms, in order to prevent their sinking.

As a remarkable example of the closing of a fissure, Humboldt mentions that, during the celebrated earthquake in 1851, in the Neapolitan province of Basilicata, a hen was found caught by both feet in the street-pavement of Barile, near Melfi.

Mr. Hopkins has very correctly shown theoretically that the fissures produced by earthquakes are very instructive as regards the formation of veins and the phenomenon of dislocation, the more recent vein displacing the older formation.

RUMBLINGS OF EARTHQUAKES.

When the great earthquake of Coseguina, in Nicaragua, took place, January 23, 1835, the subterranean noise—the sonorous waves in the earth—was heard at the same time on the island of Jamaica and on the plateau of Bogota, 8740 feet above the sea, at a greater distance than from Algiers to London. In the eruptions of the volcano on the island of St. Vincent, April 30, 1812, at 2 A.M., a noise like the report of cannons was heard, without any sensible concussion of the earth, over a space of 160,000 geographical square miles. There have also been heard subterranean thunderings for two years without earthquakes.

HOW TO MEASURE AN EARTHQUAKE-SHOCK.

A new instrument (the Seismometer) invented for this purpose by M. Kreil, of Vienna, consists of a pendulum oscillating in every direction, but unable to turn round on its point of suspension; and bearing at its extremity a cylinder, which, by means of mechanism within it, turns on its vertical axis once in twenty-four hours. Next to the pendulum stands a rod bearing a narrow elastic arm, which slightly presses the extremity of a lead-pencil against the surface of the cylinder. As long as the pendulum is quiet, the pencil traces an uninterrupted line on the surface of the cylinder; but as soon as it oscillates, this line becomes interrupted and irregular, and these irregularities indicate the time of the commencement of an earthquake, together with its duration and intensity.30

Elastic fluids are doubtless the cause of the slight and perfectly harmless trembling of the earth's surface, which has often continued for several days. The focus of this destructive agent, the seat of the moving force, lies far below the earth's surface; but we know as little of the extent of this depth as we know of the chemical nature of these vapours that are so highly compressed. At the edges of two craters,—Vesuvius and the towering rock which projects beyond the great abyss of Pichincha, near Quito,—Humboldt has felt periodic and very regular shocks of earthquakes, on each occasion from twenty to thirty seconds before the burning scoriæ or gases were erupted. The intensity of the shocks was increased in proportion to the time intervening between them, and consequently to the length of time in which the vapours were accumulating. This simple fact, which has been attested by the evidence of so many travellers, furnishes us with a general solution of the phenomenon, in showing that active volcanoes are to be considered as safety-valves for the immediate neighbourhood. There are instances in which the earth has been shaken for many successive days in the chain of the Andes, in South America. In certain districts, the inhabitants take no more notice of the number of earthquakes than we in Europe take of showers of rain; yet in such a district Bonpland and Humboldt were compelled to dismount, from the restiveness of their mules, because the earth shook in a forest for fifteen to eighteen minutes *without intermission.*

EARTHQUAKES AND THE MOON.

From a careful discussion of several thousand earthquakes which have been recorded between 1801 and 1850, and a comparison of the periods at which they occurred with the position of the moon in relation to the earth, M. Perry, of Dijon, infers that earthquakes may possibly be the result of attraction exerted by that body on the supposed fluid centre of our globe, somewhat similar to that which she exercises on the waters of the ocean; and the Committee of the Institute of France have reported favourably upon this theory.

THE GREAT EARTHQUAKE OF LISBON.

The eloquent Humboldt remarks, that the activity of an igneous mountain, however terrific and picturesque the spectacle may be which it presents to our contemplation, is always limited to a very small space. It is far otherwise with earthquakes, which, although scarcely perceptible to the eye, nevertheless simultaneously propagate their waves to a distance of many thousand miles. The great earthquake which destroyed the city of Lisbon, November 1st, 1755, was felt in the Alps, on the coast of Sweden, into the Antilles, Antigua, Barbadoes, and Martinique; in the great Canadian lakes, in Thuringia, in the flat country of northern Germany, and in the small inland lakes on the shores of the Baltic. Remote springs were interrupted in their flow,—a phenomenon attending earthquakes which had been noticed among the ancients by Demetrius the Callatian. The hot springs of Töplitz dried up and returned, inundating every thing around, and having their waters coloured with iron ochre. At Cadiz, the sea rose to an elevation of sixty-four feet; while in the Antilles, where the tide usually rises only from twenty-six to twenty-eight inches, it suddenly rose about twenty feet, the water being of an inky blackness. It has been computed that, on November 1st, 1755, a portion of the earth's surface four times greater than that of Europe was simultaneously shaken.[31] As yet there is no manifestation of force known to us (says the vivid denunciation of the philosopher), including even the murderous invention of our own race, by which a greater number of people have been killed in the short space of a few minutes: 60,000 were destroyed in Sicily in 1693, from 30,000 to 40,000 in the earthquake of Riobamba in 1797, and probably five times as many in Asia Minor and Syria under Tiberius and Justinian the elder, about the years 19 and 526.

GEOLOGICAL AGE OF THE DIAMOND.

The discovery of Diamonds in Russia, far from the tropical zone, has excited much interest among geologists. In the detritus on the banks of the Adolfskoi, no fewer than forty diamonds have been found in the gold alluvium, only twenty feet above the stratum in which the remains of mammoths and rhinoceroses are found. Hence

Humboldt has concluded that the formation of gold-veins, and consequently of diamonds, is comparatively of recent date, and scarcely anterior to the destruction of the mammoths. Sir Roderick Murchison and M. Verneuil have been led to the same result by different arguments.32

WHAT WAS ADAMANT?

Professor Tennant replies, that the Adamant described by Pliny was a sapphire, as proved by its form, and by the fact that when struck on an anvil by a hammer it would make an indentation in the metal. A true diamond, under such circumstances, would fly into a thousand pieces.

WHAT IS COAL?

The whole evidence we possess as to the nature of Coal proves it to have been originally a mass of vegetable matter. Its microscopical characters point to its having been formed on the spot in which we find it, to its being composed of vegetable tissues of various kinds, separated and changed by maceration, pressure, and chemical action, and to the introduction of its earthy matter, in a large number of instances, in a state of solution or fine molecular subdivision. Dr. Redfern, from whose communication to the British Association we quote, knows nothing to countenance the supposition that our coal-beds are mainly formed of coniferous wood, because the structures found in mother-coal, or the charcoal layer, have not the character of the glandular tissue of such wood, as has been asserted.

Geological research has shown that the immense forests from which our coal is formed teemed with life. A frog as large as an ox existed in the swamps, and the existence of insects proves that the higher order of organic creation flourished at this epoch.

It has been calculated that the available coal-beds in Lancashire amount in weight to the enormous sum of 8,400,000,000 tons. The total annual consumption of this coal, it has been estimated, amounts to 3,400,120 tons; hence it is inferred that the coal-beds of Lancashire, at the present rate of consumption, will last 2470 years. Making similar calculations for the coal-fields of South Wales, the north of England, and Scotland, it will readily be perceived how ridiculous were the forebodings which lecturing geologists delighted to indulge in a few years ago.

TORBANE-HILL COAL.

The coal of Torbane Hill, Scotland, is so highly inflammable, that it has been disputed at law whether it be true coal, or only asphaltum, or bitumen. Dr. Redfern describes it as laminated, splitting with great ease horizontally, like many cannel coals, and like them it may be lighted at a candle. In all parts of the bed stigmaria and other fossil plants occur in greater numbers than in most other coals; their distinct vascular tissue may be easily recognised by a common pocket lens, and 65½ of the mass consists of carbon.

Dr. Redfern considers that all our coals may be arranged in a scale having the Torbane-Hill coal at the top and anthracite at the bottom. Anthracite is almost pure carbon; Torbane Hill contains less fixed carbon than most other cannels: anthracite is very difficult to ignite, and gives out scarcely any gas; Torbane-Hill burns like a candle, and yields 3000 cubic feet of gas per ton, more than any other known coal, its gas being also of greatly superior illuminating power to any other. The only differences which the Torbane-Hill coal presents from others are differences of degree, not of kind. It differs from other coals in being the best gas-coal, and from other cannels in being the best cannel.

HOW MALACHITE IS FORMED.

The rich copper-ore of the Ural, which occurs in veins or masses, amid metamorphic strata associated with igneous rocks, and even in the hollows between the eruptive rocks, is worked in shafts. At the bottom of one of these, 280 feet deep, has been found an enormous irregularly-shaped botryoidal mass of *Malachite* (Greek *malache*, mountain-green), sending off strings of green copper-ore. The upper surface of it is about 18 feet long and 9 wide; and it was estimated to contain 15,000 poods, or half a million pounds, of pure and compact malachite. Sir Roderick Murchison is of opinion that this wonderful subterraneous incrustation has been produced in the stalagmitic form, during a series of ages, by copper solutions emanating from the surrounding loose and sporous mass, and trickling through it to the lowest cavity upon the subjacent solid rock. Malachite is brought chiefly from one mine in Siberia; its value as raw material is nearly one-fourth that of the same weight of pure silver, or in a manufactured state three guineas per pound avoirdupois.33

LUMPS OF GOLD IN SIBERIA.

The gold mines south of Miask are chiefly remarkable for the large lumps or *pepites* of gold which are found around the Zavod of Zarevo-Alexandroisk. Previous to 1841 were discovered here lumps of native gold; in that year a lump of twenty-four pounds was met with; and in 1843 a lump weighing about seventy-eight pounds English was found, and is now deposited with others in the Museum of the Imperial School of Mines at St. Petersburg.

SIR ISAAC NEWTON UPON BURNET'S THEORY OF THE EARTH.

In 1668, Dr. Thomas Burnet printed his *Theoria Telluris Sacra*, "an eloquent physico-theological romance," says Sir David Brewster, "which was to a certain extent adopted even by Newton, Burnet's friend. Abandoning, as some of the fathers had done, the hexaëmeron, or six days of Moses, as a physical reality, and having no knowledge of geological phenomena, he gives loose reins to his imagination, combining passages of Scripture with those of ancient authors, and presumptuously describing the future catastrophes to which the earth is to be exposed." Previous to its publication, Burnet presented a copy of his book to Newton, and requested his opinion of the theory which it propounded. Newton took "exceptions to particular passages," and a correspondence ensued. In one of Newton's letters he treats of the formation of the earth, and the other planets, out of a general chaos of the figure assumed by the earth,—of the length of the primitive days,—of the formation of hills and seas, and of the creation of the two ruling lights as the result of the clearing up of the atmosphere. He considers the account of the creation in Genesis as adapted to the judgment of the vulgar. "Had Moses," he says, "described the processes of creation as distinctly as they were in themselves, he would have made the narrative tedious and confused amongst the vulgar, and become a philosopher more than a prophet." After referring to several "causes of meteors, such as the breaking out of vapours from below, before the earth was well hardened, the settling and shrinking of the whole globe after the upper regions or surface began to be hard," Newton closes his letter with an apology for being tedious, which, he says, "he has the more reason to do, as he has not set down any thing he has well considered, or will undertake to defend."—See the Letter in the Appendix to *Sir D. Brewster's Life of Newton*, vol. ii.

The primitive condition of the earth, and its preparation for man, was a subject of general speculation at the close of the seventeenth century. Leibnitz, like his great rival (Newton), attempted to explain the formation of the earth, and of the different substances which composed it; and he had the advantage of possessing some knowledge of geological phenomena: the earth he regarded as having been originally a burning mass, whose temperature gradually diminished till the vapours were condensed into a universal ocean, which covered the highest mountains, and gradually flowed into vacuities and subterranean cavities produced by the consolidation of the earth's crust. He regarded fossils as the real remains of plants and animals which had been buried in the strata; and, in speculating on the formation of mineral substances, he speaks of crystals as the geometry of inanimate nature.—*Brewster's Life of Newton*, vol. ii. p. 100, note. (See also "The Age of the Globe," in *Things not generally Known*, p. 13.)

"THE FATHER OF ENGLISH GEOLOGY."

In 1769 was born, the son of a yeoman of Oxfordshire, William Smith. When a boy he delighted to wander in the fields, collecting "pound-stones" (*Echinites*), "pundibs" (*Terebratulæ*), and other stony curiosities; and receiving little education beyond what he taught himself, he learned nothing of classics but the name. Grown to be a man, he became a land-surveyor and civil engineer, and was much engaged in constructing canals. While thus occupied, he observed that all the rocky masses forming the substrata of the country were gently inclined to the east and south-east,—that the red sandstones and marls above the *coal-measures* passed below the beds provincially termed lias-clay and limestone—that these again passed underneath the sands, yellow limestone, and clays that form the table-land of the Coteswold Hills; while they in turn plunged beneath the great escarpment of chalk that runs from the coast of Dorsetshire northward to the Yorkshire shores of the German Ocean. He further observed that each formation of clay, sand, or limestone, held to a very great extent its own peculiar suite of fossils. The "snake-stones" (*Ammonites*) of the lias were different in form and ornament from those of the inferior oolite; and the shells of the latter, again, differed from those of the Oxford clay, Cornbrash, and Kimmeridge clay. Pondering much on these things, he came to the then unheard-of conclusion that each formation had been in its turn a sea-bottom, in the sediments of which lived and died marine animals now extinct, many specially distinctive of their own epochs in time.

Here indeed was a discovery,—made, too, by a man utterly unknown to the scientific world, and having no pretension to scientific lore. "Strata Smith's" find was unheeded for many a long year; but at length the first geologists of the day learned from the land-surveyor that superposition of strata is inseparably connected with the succession of life in time. Hooke's grand vision was at length realised, and it was indeed possible "to build up a terrestrial chronology from rotten shells" imbedded in the rocks. Meanwhile he had constructed the first geological map of England, which has served as a basis for geological maps of all other parts of the world. William Smith was now presented by the Geological Society with the Wollaston Medal, and hailed as "the Father of English Geology." He died in 1840. Till the manner as well as the fact of the first appearance of successive forms of life shall be solved, it is not easy to surmise how any discovery can be made in geology equal in value to that which we owe to the genius of William Smith.—*Saturday Review*, No. 140.

DR. BUCKLAND's GEOLOGICAL LABOURS.

Sir Henry De la Beche, in his Anniversary Address to the Geological Society in 1848, on presenting the Wollaston Medal to Dr. Buckland, felicitously observed:

It may not be generally known that, while yet a child, at your native town, Axminster in Devonshire, ammonites, obtained by your father from the lime quarries in the neighbourhood, were presented to your attention. As a scholar at Winchester, the chalk, with its flints, was brought under your observation, and there it was that your collections in natural history first began. Removed to Oxford, as a scholar of Corpus Christi College, the future teacher of geology in that University was fortunate in meeting with congenial tastes in our colleague Mr. W. J. Broderip, then a student at Oriel College. It was during your walks together to Shotover Hill, when his knowledge of conchology was so valuable to you, enabling you to distinguish the shells of the Oxford oolite, that you laid the foundation for those field-lectures, forming part of your course of geology at Oxford, which no one is likely to forget who has been so fortunate at any time as to have attended them. The fruits of your walks with Mr. Broderip formed the nucleus of that great collection, more especially remarkable for the organic remains it contains, which, after the labours of forty years, you have presented to the Geological Museum at Oxford, in grave recollection of the aid which the endowments of that University, and the leisure of its vacations, had afforded you for extensive travelling during a residence at Oxford of nearly forty-five years.

DISCOVERIES OF M. AGASSIZ.[34]

This great paleontologist, in the course of his ichthyological researches, was led to perceive that the arrangement by Cuvier according to organs did not fulfil its purpose with regard to fossil fishes, because in the lapse of ages the characteristics of their structures were destroyed. He therefore adopted the only other remaining plan, and studied the tissues, which, being less complex than the organs, are oftener found intact. The result was the very remarkable discovery, that the tegumentary membrane of fishes is so intimately connected with their organisation, that if the whole of the fish has perished except this membrane, it is practicable, by noting its characteristics, to reconstruct the animal in its most essential parts. Of the value of this principle of harmony, some idea may be formed from the circumstance, that on it Agassiz has based the whole of that celebrated classification of which he is the sole author, and by which fossil ichthyology has for the first time assumed a precise and definite shape. How essential its study is to the geologist appears from the remark of Sir Roderick Murchison, that "fossil fishes have every where proved the most exact chronometer of the age of rocks."

SUCCESSION OF LIFE IN TIME.

In the Museum of Economic Geology, in Jermyn Street, may be seen ores, metals, rocks, and whole suites of fossils stratigraphically arranged in such a manner that, with an observant eye for form, all may easily understand the more obvious scientific meanings of the Succession of Life in Time, and its bearing on geological economies. It is perhaps scarcely an exaggeration to say, that the greater number of so-called educated persons are still ignorant of the meaning of this great doctrine. They would be ashamed not to know that there are many suns and material worlds besides our own; but the science, equally grand and comprehensible, that aims at the discovery of the laws that regulated the creation, extension, decadence, and utter extinction of many successive species, genera, and whole orders of life, is ignored, or, if intruded on the attention, is looked on as an uncertain and dangerous dream,—and this in a country which was almost the nursery of geology, and which for half a century has boasted the first Geological Society in the world.—*Saturday Review*, No. 140.

PRIMITIVE DIVERSITY AND NUMBERS OF ANIMALS IN GEOLOGICAL TIMES.

Professor Agassiz considers that the very fact of certain stratified rocks, even among the oldest formations, being almost entirely made up of fragments of organised beings, should long ago have satisfied the most sceptical that both *animal and vegetable life were as active and profusely scattered upon the whole globe at all times, and during all geological periods, as they are now.* No coral reef in the Pacific contains a larger amount of organic *débris* than some of the limestone deposits of the tertiary, of the cretaceous, or of the oolitic, nay even of the paleozoic period; and the whole vegetable carpet covering the present surface of the globe, even if we were to consider only the luxuriant vegetation of the tropics, leaving entirely out of consideration the entire expanse of the ocean, as well as those tracts of land where, under less favourable circumstances, the growth of plants is more reduced,—would not form one single seam of workable coal to be compared to the many thick beds contained in the rocks of the carboniferous period alone.

ENGLAND IN THE EOCENE PERIOD.

Eocene is Sir Charles Lyell's term for the lowest group of the Tertiary system in which the dawn of recent life appears; and any one who wishes to realise what was the aspect presented by this country during the Eocene period, need only go to Sheerness. If, leaving that place behind him, he walks down the Thames, keeping close to the edge of the water, he will find whole bushels of pyritised pieces of twigs and fruits. These fruits and twigs belong to plants nearly allied to the screw-pine and custard-apple, and to various species of palms and spice-trees which now flourish in the Eastern Archipelago. At the time they were washed down from some neighbouring land, not only crocodilian reptiles, but sharks and innumerable turtles, inhabited a sea or estuary which now forms part of the London district; and huge boa-constrictors glided amongst the trees which fringed the adjoining shores.

Countless as are the ages which intervened between the Eocene period and the time when the little jawbones of Stonesfield were washed down to the place where they were to await the day when science should bring them again to light, not one mammalian genus which now lives upon our plane has been discovered amongst Eocene strata. We have existing families, but nothing more.—*Professor Owen.*

FOOD OF THE IGUANODON.

Dr. Mantell, from the examination of the anterior part of the right side of the lower jaw of an Iguanodon discovered in a quarry in Tilgate Forest, Sussex, has detected an extraordinary deviation from all known types of reptilian organisation, and which could not have been predicated; namely, that this colossal reptile, which equalled in bulk the gigantic Edentata of South America, and like them was destined to obtain support from comminuted vegetable substances, was also furnished with a large prehensile tongue and fleshy lips, to serve as instruments for seizing and cropping the foliage and branches of trees; while the arrangement of the teeth as in the ruminants, and their internal structure, which resembles that of the molars of the sloth tribe in the vascularity of the dentine, indicate adaptations for the same purpose.

Among the physiological phenomena revealed by paleontology, there is not a more remarkable one than this modification of the type of organisation peculiar to the class of reptiles to meet the conditions required by the economy of a lizard placed under similar physical relations; and destined to effect the same general purpose in the scheme of nature as the colossal Edentata of former ages and the large herbivorous mammalia of our own times.

THE PTERODACTYL—THE FLYING DRAGON.

The Tilgate beds of the Wealden series, just mentioned, have yielded numerous fragments of the most remarkable reptilian fossils yet discovered, and whose wonderful forms denote them to have thronged the shallow seas and bays and lagoons of the period. In the grounds of the Crystal Palace at Sydenham the reader will find restorations of these animals sufficiently perfect to illustrate this reptilian epoch. They include the *iguanodon*, an herbivorous lizard exceeding in size the largest elephant, and accompanied by the equally gigantic and carnivorous *megalosaurus* (great saurian), and by the two yet more curious reptiles, the *pylæosaurus* (forest, or weald, saurian) and the pterodactyl (from *pteron*, 'wing,' and *dactylus*, 'a finger'), an enormous bat-like creature, now running upon the ground like a bird; its elevated body and long neck not covered with feathers, but with skin, naked, or resplendent with glittering scales; its head like that of a lizard or crocodile, and of a size almost preposterous compared with that of the body, with its long fore extremities stretched out, and connected by a membrane with the body and hind legs.

Suddenly this mailed creature rose in the air, and realised or even surpassed in strangeness *the flying dragon of fable*: its fore-arms and its elongated wing-finger furnished with claws; hand and fingers extended, and the interspace filled up by a tough membrane; and its head and neck stretched out like that of the heron in its flight. When stationary, its wings were probably folded back like those of a bird; though perhaps, by the claws attached to its fingers, it might suspend itself from the branches of trees.

MAMMALIA IN SECONDARY ROCKS.

It was supposed till very lately that few if any Mammalia were to be found below the Tertiary rocks, *i. e.* those above the chalk; and this supposed fact was very comfortable to those who support the doctrine of "progressive development," and hold, with the notorious *Vestiges of Creation*, that a fish by mere length of time became a reptile, a lemur an ape, and finally an ape a man. But here, as in a hundred other cases, facts, when duly investigated, are against their theory. A mammal jaw had been already discovered by Mr. Brodie on the shore at the back of Swanage Point, in Dorsetshire, when Mr. Beckles, F.G.S., traced the vein from which this jaw had been procured, and found it to be a stratum about five inches thick, at the base of the Middle Purbeck beds; and after removing many thousand tons of rock, and laying bare an area of nearly 7000 square feet (the largest cutting ever made for purely scientific purposes), he found reptiles (tortoises and lizards) in hundreds; but the most important discovery was that of the jaws of at least fourteen different species of mammalia. Some of these were herbivorous, some carnivorous, connected with our modern shrews, moles, hedgehogs, &c.; but all of them perfectly developed and highly-organised quadrupeds. Ten years ago, no remains of quadrupeds were believed to exist in the Secondary strata. "Even in 1854," says Sir Charles Lyell (in a supplement to the fifth edition of his *Manual of Elementary Geology*), "only six species of mammals from rocks older than the Tertiary were known in the whole world." We now possess evidence of the existence of fourteen species, belonging to eight or nine genera, from the fresh-water strata of the Middle Purbeck Oolite. It would be rash now to fix a limit in past time to the existence of quadrupeds.—*The Rev. C. Kingsley.*

FOSSIL HUMAN BONES.

In the paleontological collection in the British Museum is preserved a considerable portion of a human skeleton imbedded in a slab of rock, brought from Guadaloupe, and often referred to in opposition to the statement that hitherto *no fossil human hones have been found.* The presence of these bones, however, has been explained by the circumstance of a battle and the massacre of a tribe of Galtibis by the Caribs, which took place near the spot in which the bones were found about 130 years ago; for as the bodies of the slain were interred on the seashore, their skeletons may have been subsequently covered by sand-drift, which has since consolidated into limestone.

It will be seen by reference to the *Philosophical Transactions*, that on the reading of the paper upon this discovery to the Royal Society, in 1814, Sir Joseph Banks, the president, considered the "fossil" to be of very modern formation, and that probably, from the contiguity of a volcano, the temperature of the water may have been raised at some time, and dissolving carbonate of lime readily, may have deposited about the skeleton in a comparatively short period hard and solid stone. Every person may be convinced of the rapidity of the formation and of the hardness of such stone by inspecting the inside of tea-kettles in which hard water is boiled.

Descriptions of petrifactions of human bodies appear to refer to the conversion of bodies into adipocere, and not into stone. All the supposed cases of petrifaction are probably of this nature. The change occurs only when the coffin becomes filled with water. The body, converted into adipocere, floats on the water. The supposed cases of changes of position in the grave, bursting open the coffin-lids, turning over, crossing of limbs, &c., formerly attributed to the coming to life of persons buried who were not dead, is now ascertained to be due to the same cause. The chemical change into adipocere, and the evolution of gases, produce these movements of dead bodies.—*Mr. Trail Green.*

THE MOST ANCIENT FISHES.

Among the important results of Sir Roderick Murchison's establishment of the Silurian system is the following:

That as the Lower Silurian group, often of vast dimensions, has never afforded the smallest vestige of a Fish, though it abounds in numerous species of the *marine* classes,—corals, *crinoidea, mollusca,* and *crustacea*; and as in Scandinavia and Russia, where it is based on rocks void of fossils, its lowest stratum contains *fucoids* only,—Sir R. Murchison has, after fifteen years of laborious research steadily directed to this point, arrived at the conclusion, that a very long period elapsed after life was breathed into the waters before the lowest order of vertebrata was created; the earliest fishes being those of the Upper Silurian rocks, which he was the first to discover, and which he described "as the most ancient beings of their class which have yet been brought to light." Though the Lower Silurian rocks of various parts of the world have since been ransacked by multitudes of prying geologists, who have exhumed from them myriads of marine fossils, not a single ichthyolite has been found in any stratum of higher antiquity than the Upper Silurian group of Murchison.

The most remarkable of all fossil fishes yet discovered have been found in the Old Red Sandstone cliffs at Dorpat, where the remains are so gigantic (one bone measuring *two feet nine inches* in length) that they were at first supposed to belong to saurians.

Sir Roderick's examination of Russia has, in short, proved that *the ichthyolites and mollusks which, in Western Europe, are separately peculiar to smaller detached basins, were here (in the British Isles) cohabitants of many parts of the same great sea.*

EXTINCT CARNIVOROUS ANIMALS OF BRITAIN.

Professor Owen has thus forcibly illustrated the Carnivorous Animals which preyed upon and restrained the undue multiplication of the vegetable feeders. First we have the bear family, which is now represented in this country only by the badger. We were once blest, however, with many bears. One species seems to have been identical with the existing brown bear of the European continent. Far larger and more formidable was the gigantic cave-bear (*Ursus spelæus*), which surpassed in size his grisly brother of North America. The skull of the cave-bear differs very much in shape from that of its small brown relative just alluded to; the forehead, in particular, is much higher,—to be accounted for by an arrangement of air-cells similar to those which we have already remarked in the elephant. The cave-bear has left its remains in vast abundance in Germany. In our own caves, the bones of hyænas are found in greater quantities. The marks which the teeth of the hyæna make upon the bones which it gnaws are quite unmistakable. Our English hyænas had the most undiscriminating appetite, preying upon every creature, their own species amongst others. Wolves, not distinguishable from those which now exist in France and Germany, seem to have kept company with the hyænas; and the *Felis spelæa*, a sort of lion, but larger than any which now exists, ruled over all weaker brutes. Here, says Professor Owen, we have the original British Lion. A species of *Machairodus* has left its remains at Kent's Hole, near Torquay. In England we had also the beaver, which still lingers on the Danube and the Rhone, and a larger species, which has been called Trogontherium (gnawing beast), and a gigantic mole.

THE GREAT CAVE TIGER OR LION OF BRITAIN.

Remains of this remarkable animal of the drift or gravel period of this country have been found at Brentford and elsewhere near London. Speaking of this animal, Professor Owen observes, that "it is commonly supposed that the Lion, the Tiger, and the Jaguar are animals peculiarly adapted to a tropical climate. The genus Felis (to which these animals belong) is, however, represented by specimens in high northern latitudes, and in all the intermediate countries to the equator." The chief condition necessary for the presence of such animals is an abundance of the vegetable-feeding animals. It is thus that the Indian tiger has been known to follow the herds of antelope and deer in the lofty mountains of the Himalaya to the verge of perpetual snow, and far into Siberia. "It need not, therefore," continues Professor Owen, "excite surprise that indications should have been discovered in the fossil relics of the ancient mammalian population of Europe of a large feline animal, the contemporary of the mammoth, of the tichorrhine rhinoceros, of the great gigantic cave-bear and hyæna, and the slayer of the oxen, deer, and equine quadrupeds that so abounded during the same epoch." The dimensions of this extinct animal equal those of the largest African lion or Bengal tiger; and some bones have been found which seem to imply that it had even more powerful limbs and larger paws.

THE MAMMOTHS OF THE BRITISH ISLES.

Dr. Buckland has shown that for long ages many species of carnivorous animals now extinct inhabited the caves of the British islands. In low tracts of Yorkshire, where tranquil lacustrine (lake-like) deposits have occurred, bones (even those of the lion) have been found so perfectly unbroken and unworn, in fine gravel (as at Market Weighton), that few persons would be disposed to deny that such feline and other animals once roamed over the British isles, as well as other European countries. Why, then, is it improbable that large elephants, with a peculiarly thick integument, a close coating of wool, and much long shaggy hair, should have been the occupants of wide tracts of Northern Europe and Asia? This coating, Dr. Fleming has well remarked, was probably as impenetrable to rain and cold as that of the monster ox of the polar circle. Such is the opinion of Sir Roderick Murchison, who thus accounts for the disappearance of the mammoths from Britain:

When we turn from the great Siberian continent, which, anterior to its elevation, was the chief abode of the mammoths, and look to the other parts of Europe, where their remains also occur, how remarkable is it that we find the number of these creatures to be justly proportionate to the magnitude of the ancient masses of land which the labours of geologists have defined! Take the British isles, for example, and let all their low, recently elevated districts be submerged; let, in short, England be viewed as the comparatively small island she was when the ancient estuary of the Thames, including the plains of Hyde Park, Chelsea, Hounslow, and Uxbridge, were under the water; when the Severn extended far into the heart of the kingdom, and large eastern tracts of the island were submerged,—and there will then remain but moderately-sized feeding-grounds for the great quadrupeds whose bones are found in the gravel of the adjacent rivers and estuaries.

This limited area of subsistence could necessarily only keep up a small stock of such animals; and, just as we might expect, the remains of British mammoths occur in very small numbers indeed, when compared with those of the great charnel-houses of Siberia, into which their bones had been carried down through countless ages from the largest mass of surface which geological inquiries have yet shown to have been *dry land* during that epoch.

The remains of the mammoth, says Professor Owen, have been found in all, or almost all, the counties of England. Off the coast of Norfolk they are met with in vast abundance. The fishermen who go to catch turbot between the mouth of the Thames and the Dutch coast constantly get their nets entangled in the tusks of the mammoth. A collection of tusks and other remains, obtained in this way, is to be seen at

Ramsgate. In North America, this gigantic extinct elephant must have been very common; and a large portion of the ivory which supplies the markets of Europe is derived from the vast mammoth graveyards of Siberia.

The mammoth ranged at least as far north as 60°. There is no doubt that, at the present day, many specimens of the musk-ox are annually becoming imbedded in the mud and ice of the North-American rivers.

It is curious to observe, that the mammoth teeth which are met with in caves generally belonged to young mammoths, who probably resorted thither for shelter before increasing age and strength emboldened them to wander far afield.

THE RHINOCEROS AND HIPPOPOTAMUS OF ENGLAND.

The mammoth was not the only giant that inhabited England in the Pliocene or Upper Tertiary period. We had also here the *Rhinoceros tichorrhinus*, or "strongly walled about the nose," remains of which have been discovered in enormous quantities in the brickfields about London. Pallas describes an entire specimen of this creature, which was found near Yakutsk, the coldest town on the globe. Another rhinoceros, *leptorrhinus* (fine nose), dwelt with the elephant of Southern Europe. In Siberia has been discovered the Elaimotherium, forming a link between the rhinoceros and the horse.

In the days of the mammoth, we had also in England a Hippopotamus, rather larger than the species which now inhabits the Nile. Of our British hippopotamus some remains were dug up by the workmen in preparing the foundations of the New Junior United Service Club-house, in Regent-street.

THE ELEPHANT AND TORTOISE.

The idea of an Elephant standing on the back of a Tortoise was often laughed at as an absurdity, until Captain Cautley and Dr. Falconer at length discovered in the hills of Asia the remains of a tortoise in a fossil state of such a size that an elephant could easily have performed the above feat.

COEXISTENCE OF MAN AND THE MASTODON.

Dr. C. F. Winslow has communicated to the Boston Society of Natural History the discovery of the fragment of a human cranium 180 feet below the surface of the Table Mountain, California. Now the mastodon's bones being found in the same deposits, points very clearly to the probability of the appearance of the human race on the western portions of North America at least before the extinction of those huge creatures. Fragments of mastodon and *Elephas primigenius* have been taken ten and twenty feet below the surface in the above locality; where this discovery of human and mastodon remains gives strength to the possible truth of an old Indian tradition,— the contemporary existence of the mammoth and aboriginals in this region of the globe.

HABITS OF THE MEGATHERIUM.

Much uncertainty has been felt about the habits of the Megatherium, or Great Beast. It has been asked whether it burrowed or climbed, or what it did; and difficulties have presented themselves on all sides of the question. Some have thought that it lived in trees as much larger than those which now exist as the Megatherium itself is larger than the common sloth.35 This, however, is now known to be a mistake. It did not climb trees—it pulled them down; and in order to do this the hinder parts of its skeleton were made enormously strong, and its prehensile fore-legs formed so as to give it a tremendous power over any thing which it grasped. Dr. Buckland suggested that animals which got their living in this way had a very fair chance of having their heads broken. While Professor Owen was still pondering over this difficulty, the skull of a cognate animal, the Mylodon, came into his hands. Great was his delight when he found that the mylodon not only had his head broken, but broken in two different places, at two different times; and moreover so broken that the injury could only have been inflicted by some such agent as a fallen tree. The creature had recovered from the first blow, but had evidently died of the second. This tribe had, as it turns out, two skulls, an outer and an inner one—given them, as it would appear, expressly with a view to the very dangerous method in which they were intended to obtain their necessary food.

The dentition of the megatherium is curious. The elephant gets teeth as he wants them. Nature provided for the comfort of the megatherium in another way. It did not get new teeth, but the old ones went on for ever growing as long as the animal lived; so that as fast as one grinding surface became useless, another supplied its place.

THE DINOTHERIUM, OR TERRIBLE BEAST.

The family of herbivorous Cetaceans are connected with the Pachydermata of the land by one of the most wonderful of all the extinct creatures with which geologists have made us acquainted. This is the *Dinotherium*, or Terrible Beast. The remains of this animal were found in Miocene sands at Eppelsheim, about forty miles from Darmstadt. It must have been larger than the largest extinct or living elephant. The most remarkable peculiarity of its structure is the enormous tusks, curving downwards and terminating its lower jaw. It appears to have lived in the water, where the immense weight of these formidable appendages would not be so inconvenient as on land. What these tusks were used for is a mystery; but perhaps they acted as pickaxes in digging up trees and shrubs, or as harrows in raking the bottom of the water. Dr. Buckland used to suggest that they were perhaps employed as anchors, by means of which the monster might fasten itself to the bank of a stream and enjoy a comfortable nap. The extreme length of the *Dinotherium* was about eighteen feet. Professor Kemp, in his restoration of the animal, has given it a trunk like that of the elephant, but not so long, and the general form of the tapir.—*Professor Owen.*

THE GLYPTODON.

There are few creatures which we should less have expected to find represented in fossil history by a race of gigantic brethren than the armadillo. The creature is so small, not only in size but in all its works and ways, that we with difficulty associate it with the idea of magnitude. Yet Sir Woodbine Parish has discovered evidences of enormous animals of this family having once dwelt in South America. The huge loricated (plated over) creature whose relics were first sent has received the name of Glyptodon, from its sculptured teeth. Unlike the small armadillos, it was unable to roll itself up into a ball; though an enormous carnivore which lived in those days must have made it sometimes wish it had the power to do so. When attacked, it must have crouched down, and endeavoured to make its huge shell as good a defence as possible.—*Professor Owen.*

INMATES OF AN AUSTRALIAN CAVERN.

From the fossil-bone caverns in Wellington Valley, in 1830, were sent to Professor Owen several bones which belonged, as it turned out, to gigantic kangaroos, immensely larger than any existing species; to a kind of wombat, to formidable dasyures, and several other genera. It also appeared that the bones, which were those of herbivores, had evidently belonged to young animals, while those of the carnivores were full-sized; a fact which points to the relations between the two families having been any thing but agreeable to the herbivores.

THE POUCH-LION OF AUSTRALIA.

The *Thylacoleo* (Pouch-Lion) was a gigantic marsupial carnivore, whose character and affinities Professor Owen has, with exquisite scientific tact, made out from very small indications. This monster, which had kangaroos with heads three feet long to feed on, must have been one of the most extraordinary animals of the antique world.

THE CONEY OF SCRIPTURE.

Paleontologists have pointed out the curious fact that the Hyrax, called 'coney' in our authorised version of the Bible, is really only a diminutive and hornless rhinoceros. Remains have been found at Eppelsheim which indicate an animal more like a gigantic Hyrax than any of the existing rhinoceroses. To this the name of *Acerotherium* (Hornless Beast) has been given.

A THREE-HOOFED HORSE.

Professor Owen describes the *Hipparion*, or Three-hoofed Horse, as the first representative of a family so useful to mankind. This animal, in addition to its true hoof, appears to have had two additional elementary hoofs, analogous to those which we see in the ox. The object of these no doubt was to enable the Hipparion to extricate his foot with greater ease than he otherwise could when it sank through the swampy ground on which he lived.

TWO MONSTER CARNIVORES OF FRANCE.

A huge carnivorous creature has been found in Miocene strata in France, in which country it preyed upon the gazelle and antelope. It must have been as large as a grisly bear, but in general appearance and teeth more like a gigantic dog. Hence the name of *Amphicyon* (Doubtful Dog) has been assigned to it. This animal must have derived part of its support from vegetables. Not so the coeval monster which has been called *Machairodus* (Sabre-tooth). It must have been somewhat akin to the tiger, and is by far the most formidable animal which we have met with in our ascending progress through the extinct mammalia.—*Professor Owen.*

GEOLOGY OF THE SHEEP.

No unequivocal fossil remains of the sheep have yet been found in the bone-caves, the drift, or the more tranquil stratified newer Pliocene deposits, so associated with the fossil bones of oxen, wild-boars, wolves, foxes, otters, &c., as to indicate the coevality of the sheep with those species, or in such an altered state as to indicate them to have been of equal antiquity. Professor Owen had his attention particularly directed to this point in collecting evidence for a history of British Fossil Mammalia. No fossil core-horns of the sheep have yet been any where discovered; and so far as this negative evidence goes, we may infer that the sheep is not geologically more ancient than man;

that it is not a native of Europe, but has been introduced by the tribes who carried hither the germs of civilisation in their migrations westward from Asia.

THE TRILOBITE.

Among the earliest races we have those remarkable forms, the Trilobites, inhabiting the ancient ocean. These crustacea remotely resemble the common wood-louse, and like that animal they had the power of rolling themselves into a ball when attacked by an enemy. The eye of the trilobite is a most remarkable organ; and in that of one species, *Phacops caudatus*, not less than 250 lenses have been discovered. This remarkable optical instrument indicates that these creatures lived under similar conditions to those which surround the crustacea of the present day.—*Hunt's Poetry of Science.*

PROFITABLE SCIENCE.

In that strip of reddish colour which runs along the cliffs of Suffolk, and is called the Redcrag, immense quantities of cetacean remains have been found. Four different kinds of whales, little inferior in size to the whalebone whale, have left their bones in this vast charnel-house. In 1840, a singularly perplexing fossil was brought to Professor Owen from this Redcrag. No one could say what it was. He determined it to be the tooth of a cetacean, a unique specimen. Now the remains of cetaceans in the Suffolk crag have been discovered in such enormous quantities, that many thousands a-year are made by converting them into manure.

EXTINCT GIGANTIC BIRDS OF NEW ZEALAND.

In the islands of New Zealand have been found the bones of large extinct wingless Birds, belonging to the Post Tertiary or Recent system, which have been deposited by the action of rivers. The bird is named *Moa* by the natives, and *Dinornis* by naturalists: some of the bones have been found in two caves in the North Island, and have been sold by the natives at an extraordinary price. The caves occur in limestone rocks, and the bones are found beneath earth and a soft deposit of carbonate of lime. The largest of the birds is stated to have stood thirteen or fourteen feet, or twice the height of the ostrich; and its egg large enough to fill the hat of a man as a cup. Several statements have appeared of these birds being still in existence, but there is every reason to believe the Moa to be altogether extinct.

An extensive collection of remains of these great wingless birds has been collected in New Zealand by Mr. Walter Mantell, and deposited in the British Museum. Among these bones Professor Owen has discovered a species which he regards as the most remarkable of the feathered class for its prodigious strength and massive proportions, and which he names *Dinornis elephantopus*, or elephant-footed, of which the Professor has been able to construct an entire lower limb: the length of the metatarsal bone is 9¼ inches, the breadth of the lower end being 5-1/3 inches. The extraordinary proportions of the metatarsus of this wingless bird will, however, be still better understood by comparison with the same bone in the ostrich, in which the metatarsus is 19 inches in length, the breadth of its lower end being only 2½ inches. From the materials accumulated by Mr. Mantell, the entire skeleton of the *Dinornis elephantopus* has been reconstructed; and now forms a worthy companion of the Megatherium and Mastodon in the gallery of fossil remains in the British Museum. This species of *Dinornis* appears to have been restricted to the Middle Island of New Zealand.36

Another specimen of the remains of the *Dinornis* is preserved in the Museum of the Royal College of Surgeons, in Lincoln's-Inn Fields; and the means by which the college obtained this valuable acquisition is thus graphically narrated by Mr. Samuel Warren, F.R.S.:

In the year 1839, Professor Owen was sitting alone in his study, when a shabbily-dressed man made his appearance, announcing that he had got a great curiosity, which he had brought from New Zealand, and wished to dispose of to him. It had the appearance of an old marrow-bone, about six inches in length, and rather more than two inches in thickness, *with both extremities broken off*, and Professor Owen considered that, to whatever animal it might have belonged, the fragment must have lain in the earth for centuries. At first he considered this same marrow-bone to have belonged to an ox, at all events to a quadruped; for the wall or rim of the bone was six times as thick as the bone of any bird, even of the ostrich. He compared it with the bones in the skeleton of an ox, a horse, a camel, a tapir, and every quadruped apparently possessing a bone of that size and configuration; but it corresponded with none. On this he very narrowly examined the surface of the bony rim, and at length became satisfied that this fragment must have belonged to *a bird*!—to one at least as large as an ostrich, but of a totally different species; and consequently one never before heard of, as an ostrich was by far the biggest bird known.

From the difference in the *strength* of the bone, the ostrich being unable to fly, so must have been unable this unknown bird; and so our anatomist came to the conclusion that this old shapeless bone indicated the former existence in New Zealand of some huge bird, at least as great as an ostrich, but of a far heavier and more sluggish kind. Professor Owen was confident of the validity of his conclusions, but would communicate that confidence to no one else; and notwithstanding attempts to dissuade him from committing his views to the public, he printed his deductions in the *Transactions of the Zoological Society for 1839*, where fortunately they remain on record as conclusive evidence of the fact of his having then made this guess, so to speak, in the dark. He caused the bone, however, to be engraved; and having sent a hundred copies of the engraving to New Zealand, in the hope of their being distributed and leading to interesting results, he patiently waited for three years,—viz. till the year 1842,—when he received intelligence from Dr. Buckland, at Oxford, that a great box, just arrived from New Zealand, consigned to himself, was on its way, unopened, to Professor Owen, who found it filled with bones, palpably of a bird, one of which bones was three feet in length, and much more than double the size of any bone in the ostrich!

And out of the contents of this box the Professor was positively enabled to articulate almost the entire skeleton of a huge wingless bird between TEN and ELEVEN feet in height, its bony structure in strict conformity with the fragment in question; and that skeleton may at any time be seen at the Museum of the College of Surgeons, towering over, and nearly twice the height of, the skeleton of an ostrich; and at its feet lying the old bone from which alone consummate anatomical science had deduced such an astounding reality,—the existence of an enormous extinct creature of the bird kind, in an island where previously no bird had been known to exist larger than a pheasant or a common fowl!—*Lecture on the Moral and Intellectual Development of the present Age.*[37]

"THE MAESTRICHT SAURIAN FOSSIL" A FRAUD.

In 1795, there was stated to have been discovered in the stone quarries adjoining Maestricht the remains of the gigantic *Moscæsaurus* (Saurian of the Meuse), an aquatic reptile about twenty-five feet long, holding an intermediate place between the Monitors and Iguanas. It appears to have had webbed feet, and a tail of such construction as to have served for a powerful oar, and enabled the animal to stem the waves of the ocean, of which Cuvier supposed it to have been an inhabitant. It is thus referred to by Dr. Mantell, in his *Medals of Creation*: "A specimen, with the jaws and bones of the palate, now in the Museum at Paris, has long been celebrated; and is still the most precious relic of this extinct reptile hitherto discovered." An admirable cast of this specimen is preserved in the British Museum, in a case near the bones of the Iguanodon. This is, however, useless, as Cuvier is proved to have been imposed upon in the matter.

M. Schlegel has reported to the French Academy of Sciences, that he has ascertained beyond all doubt that the famous fossil saurian of the quarries of Maestricht, described as a wonderful curiosity by Cuvier, is nothing more than an impudent fraud. Some bold impostor, it seems, in order to make money, placed a quantity of bones in the quarries in such a way as to give them the appearance of having been recently dug up, and then passed them off as specimens of antediluvian creation. Being successful in this, he went the length of arranging a number of bones so as to represent an entire skeleton; and had thus deceived the learned Cuvier. In extenuation of Cuvier's credulity, it is stated that the bones were so skilfully coloured as to make them look of immense antiquity, and he was not allowed to touch them lest they should crumble to pieces. But when M. Schlegel subjected them to rude handling, he found that they were comparatively modern, and that they were placed one by the other without that profound knowledge of anatomy which was to have been expected from the man bold enough to execute such an audacious fraud.

"THE OLDEST PIECE OF WOOD UPON EARTH."

The most remarkable vegetable relic which the Lower Old Red Sandstone has given us is a small fragment of a coniferous tree of the Araucarian family, which formed one of the chief ornaments of the late Hugh Miller's museum, and to which he used to point as the oldest piece of wood upon earth. He found it in one of the ichthyolite beds of Cromarty, and thus refers to it in his *Testimony of the Rocks*:

On what perished land of the early paleozoic ages did this venerably antique tree cast root and flourish, when the extinct genera Pterichthys and Coccoeteus were enjoying life by millions in the surrounding seas, long ere the flora or fauna of the coal measures had begun to be?

The same nodule which enclosed this lignite contained part of another fossil, the well-marked scales of *Diplacanthus striatus*, an ichthyolite restricted to the Lower Old Red Sandstone exclusively. If there be any value in paleontological evidence, this Cromarty lignite must have been deposited in a sea inhabited by the Coccoeteus and Diplacanthus. It is demonstrable that, while yet in a recent state, a Diplacanthus lay down and died beside it; and the evidence in the case is unequivocally this, that in the oldest portion of the oldest terrestrial flora yet known there occurs the fragment of a tree quite as high in the scale as the stately Norfolk-Island pine or the noble cedar of Lebanon.

NO FOSSIL ROSE.

Professor Agassiz, in a lecture upon the trees of America, states a remarkable fact in regard to the family of the rose,—which includes among its varieties not only many of the most beautiful flowers, but also the richest fruits, as the apple, pear, peach, plum, apricot, cherry, strawberry, raspberry, &c.,—namely, that *no fossil plants belonging to this family have ever been discovered by geologists*! This M. Agassiz regards as conclusive evidence that the introduction of this family of plants upon the earth was coeval with, or subsequent to, the creation of man, to whose comfort and happiness they seem especially designed by a wise Providence to contribute.

CHANGES ON THE EARTH'S SURFACE.

In the Imperial Library at Paris is preserved a manuscript work by an Arabian writer, Mohammed Karurini, who flourished in the seventh century of the Hegira, or at the close of the thirteenth century of our era. Herein we find several curious remarks on aerolites and earthquakes, and the successive changes of position which the land and sea have undergone. Of the latter class is the following beautiful passage from the narrative of Khidz, an allegorical personage:

I passed one day by a very ancient and wonderfully populous city, and asked one of its inhabitants how long it had been founded. "It is indeed a mighty city," replied he; "we know not how long it has existed, and our ancestors were on this subject as ignorant as ourselves." Five centuries afterwards, as I passed by the same place, I could not perceive the slightest vestige of the city. I demanded of a peasant who was gathering herbs upon its former site how long it had been destroyed. "In sooth, a strange question," replied he; "the ground here has never been different from what you now behold it." "Was there not of old," said I, "a splendid city here?" "Never," answered he, "so far as we have seen; and never did our fathers speak to us of any such." On my return there five hundred years afterwards, *I found the sea in the same place*; and on its shores were a party of fishermen, of whom I inquired how long the land had been covered by the waters. "Is this a question," say they, "for a man like you? This spot has always been what it is now." I again returned five hundred years afterwards; the sea had disappeared: I inquired of a man who stood alone upon the spot how long this change had taken place, and he gave me the same answer as I had received before. Lastly, on coming back again after an equal lapse of time, I found there a flourishing city, more populous and more rich in beautiful buildings than the city I had seen the first time; and when I would fain have informed myself concerning its origin, the inhabitants answered me, "Its rise is lost in remote antiquity: we are ignorant how long it has existed, and our fathers were on this subject as ignorant as ourselves."

This striking passage was quoted in the *Examiner*, in 1834. Surely in this fragment of antiquity we trace the "geological changes" of modern science.

GEOLOGICAL TIME.

Many ingenious calculations have been made to approximate the dates of certain geological events; but these, it must be confessed, are more amusing than instructive. For example, so many inches of silt are yearly laid down in the delta of the

Mississippi—how many centuries will it have taken to accumulate a thickness of 30, 60, or 100 feet? Again, the ledges of Niagara are wasting at the rate of so many feet per century—how many years must the river have taken to cut its way back from Queenstown to the present Falls? Again, lavas and melted basalts cool, according to the size of the mass, at the rate of so many degrees in a given time—how many millions of years must have elapsed, supposing an original igneous condition of the earth, before its crust had attained a state of solidity? or further, before its surface had cooled down to the present mean temperature? For these and similar computations, the student will at once perceive we want the necessary uniformity of factor; and until we can bring elements of calculation as exact as those of astronomy to bear on geological chronology, it will be better to regard our "eras" and "epochs" and "systems" as so many terms, indefinite in their duration, but sufficient for the magnitude of the operations embraced within their limits.—*Advanced Textbook of Geology, by David Page, F.G.S.*

M. Rozet, in 1841, called attention to the fact, that the causes which have produced irregularities in the structure of the globe have not yet ceased to act, as is proved by earthquakes, volcanic eruptions, slow and continuous movements of the crust of the earth in certain regions, &c. We may, therefore, yet see repeated the great catastrophes which the surface of the earth has undergone anteriorly to the historical period.

At the meeting of the British Association in 1855, Mr. Hopkins excited much controversy by his startling speculation—that 9000 years ago the site on which London now stands was in the torrid zone; and that, according to perpetual changes in progress, the whole of England would in time arrive within the Arctic circle.

CURIOUS CAUSE OF CHANGE OF LEVEL.

Professor Hennessey, in 1857, *found the entire mass of rock and hill on which the Armagh Observatory is erected to be slightly, but to an astronomer quite perceptibly, tilted or canted, at one season to the east, at another to the west.* This he at first attributed to the varying power of the sun's radiation to heat and expand the rock throughout the year; but he subsequently had reason to attribute it rather to the infiltration of water to the parts where the clay-slate and limestone rocks met, the varying quantity of the water exerting a powerful hydrostatic energy by which the position of the rock is slightly varied.

Now Armagh and its observatory stand at the junction of the mountain limestone with the clay-slate, having, as it were, one leg on the former and the other on the latter; and both rocks probably reach downwards 1000 or 2000 feet. When rain falls, the one will absorb more water than the other; both will gain an increase of conductive power; but the one which has absorbed most water will have the greatest increase, and being thus the better conductor, will *draw a greater portion of heat from the hot nucleus below to the surface*—will become, in fact, temporarily hotter, and, as a consequence, *expand more than the other*. In a word, *both rocks will expand at the wet season; but the best conductor, or most absorbent rock, will expand most, and seem to tilt the hill to one side; at the dry season it will subside most, and the hill will seem to be tilted in the opposite direction.*

The fact is curious, and not less so are the results deducible from it. First, hills are higher at one season than another; a fact we might have supposed, but never could have ascertained by measurement. Secondly, they are highest, not, as we should have supposed, at the hottest season, but at the wettest. Thirdly, it is from the *different rates* of expansion of different rocks that this has been discovered. Fourthly, it is by converse with the *heavens* that it has been made known to us. A variation of probably half a second, or less, in the right ascension of three or four stars, observed at different seasons, no doubt revealed the fact to the sagacious astronomer of Armagh, and even enabled him to divine its cause.

Professor Hennessey observes in connection with this phenomenon, that a very small change of ellipticity would suffice to lay bare or submerge extensive tracts of the globe. If, for example, the mean ellipticity of the ocean increased from 1/300 to 1/299, the level of the sea would be raised at the equator by about 228 feet, while under the parallel of 52° it would be depressed by 196 feet. Shallow seas and banks in the latitudes of the British isles, and between them and the pole, would thus be converted into dry land, while low-lying plains and islands near the equator would be submerged. If similar phenomena occurred during early periods of geological history, they would manifestly influence the distribution of land and water during these periods; and with such a direction of the forces as that referred to, they would tend to increase the proportion of land in the polar and temperate regions of the earth, as compared with the equatorial regions during successive geological epochs. Such maps as those published by Sir Charles Lyell on the distribution of land and water in Europe during the Tertiary period, and those of M. Elie de Beaumont, contained in Beaudant's *Geology*, would, if sufficiently extended, assist in verifying or disproving these views.

THE OUTLINES OF CONTINENTS NOT FIXED.

Continents (says M. Agassiz) are only a patchwork formed by the emergence and subsidence of land. These processes are still going on in various parts of the globe. Where the shores of the continent are abrupt and high, the effect produced may be slight, as in Norway and Sweden, where a gradual elevation is going on without much alteration in their outlines. But if the continent of North America were to be depressed

1000 feet, nothing would remain of it except a few islands, and any elevation would add vast tracts to its shores.

The west of Asia, comprising Palestine and the country about Ararat and the Caspian Sea, is below the level of the ocean, and a rent in the mountain-chains by which it is surrounded would transform it into a vast gulf.

Meteorological Phenomena.

THE ATMOSPHERE.

A philosopher of the East, with a richness of imagery truly oriental, describes the Atmosphere as "a spherical shell which surrounds our planet to a depth which is unknown to us, by reason of its growing tenuity, as it is released from the pressure of its own superincumbent mass. Its upper surface cannot be nearer to us than 50, and can scarcely be more remote than 500, miles. It surrounds us on all sides, yet we see it not; it presses on us with a load of fifteen pounds on every square inch of surface of our bodies, or from seventy to one hundred tons on us in all, yet we do not so much as feel its weight. Softer than the softest down, more impalpable than the finest gossamer, it leaves the cobweb undisturbed, and scarcely stirs the lightest flower that feeds on the dew it supplies; yet it bears the fleets of nations on its wings around the world, and crushes the most refractory substances with its weight. When in motion, its force is sufficient to level the most stately forests and stable buildings with the earth—to raise the waters of the ocean into ridges like mountains, and dash the strongest ships to pieces like toys. It warms and cools by turns the earth and the living creatures that inhabit it. It draws up vapours from the sea and land, retains them dissolved in itself or suspended in cisterns of clouds, and throws them down again as rain or dew when they are required. It bends the rays of the sun from their path to give us the twilight of evening and of dawn; it disperses and refracts their various tints to beautify the approach and the retreat of the orb of day. But for the atmosphere sunshine would burst on us and fail us at once, and at once remove us from midnight darkness to the blaze of noon. We should have no twilight to soften and beautify the landscape; no clouds to shade us from the searching heat; but the bald earth, as it revolved on its axis, would turn its tanned and weakened front to the full and unmitigated rays of the lord of day. It affords the gas which vivifies and warms our frames, and receives into itself that which has been polluted by use and is thrown off as noxious. It feeds the flames of life exactly as it does that of the fire—it is in both cases consumed and affords the food of consumption—in both cases it becomes combined with charcoal, which requires it for combustion and is removed by it when this is over."

UNIVERSALITY OF THE ATMOSPHERE.

It is only the girdling, encircling air that flows above and around all that makes the whole world kin. The carbonic acid with which to-day our breathing fills the air, to-morrow makes its way round the world. The date-trees that grow round the falls of the Nile will drink it in by their leaves; the cedars of Lebanon will take of it to add to their stature; the cocoa-nuts of Tahiti will grow rapidly upon it; and the palms and bananas of Japan will change it into flowers. The oxygen we are breathing was distilled for us some short time ago by the magnolias of the Susquehanna; the great trees that skirt the Orinoco and the Amazon, the giant rhododendrons of the Himalayas, contributed to it, and the roses and myrtles of Cashmere, the cinnamon-tree of Ceylon, and the forest, older than the Flood, buried deep in the heart of Africa, far behind the Mountains of the Moon. The rain we see descending was thawed for us out of the icebergs which have watched the polar star for ages; and the lotus-lilies have soaked up from the Nile, and exhaled as vapour, snows that rested on the summits of the Alps.—*North-British Review.*

THE HEIGHT OF THE ATMOSPHERE.

The differences existing between that which appertains to the air of heaven (the realms of universal space) and that which belongs to the strata of our terrestrial atmosphere are very striking. It is not possible, as well-attested facts prove, perfectly to explain the operations at work in the much-contested upper boundaries of our atmosphere. The extraordinary lightness of whole nights in the year 1831, during which small print might be read at midnight in the latitudes of Italy and the north of Germany, is a fact directly at variance with all we know according to the researches on the crepuscular theory and the height of the atmosphere. The phenomena of light depend upon conditions still less understood; and their variability at twilight, as well as in the zodiacal light, excite our astonishment. Yet the atmosphere which surrounds the earth is not thicker in proportion to the bulk of our globe than the line of a circle two inches in diameter when compared with the space which it encloses, or the down on the skin of a peach in comparison with the fruit inside.

COLOURS OF THE ATMOSPHERE.

Pure air is blue, because, according to Newton, the molecules of the air have the thickness necessary to reflect blue rays. When the sky is not perfectly pure, and the atmosphere is blended with perceptible vapours, the diffused light is mixed with a large proportion of white. As the moon is yellow, the blue of the air assumes somewhat of a greenish tinge, or, in other words, becomes blended with yellow.— *Letter from Arago to Humboldt*; *Cosmos*, vol. iii.

BEAUTY OF TWILIGHT.

This phenomenon is caused by the refraction of solar light enabling it to diffuse itself gradually over our hemisphere, obscured by the shades of night, long before the sun appears, even when that luminary is eighteen degrees below our horizon. It is towards the poles that this reflected splendour of the great luminary is longest visible, often changing the whole of the night into a magic day, of which the inhabitants of southern Europe can form no adequate conception.

HOW PASCAL WEIGHED THE ATMOSPHERE.

Pascal's treatise on the weight of the whole mass of air forms the basis of the modern science of Pneumatics. In order to prove that the mass of air presses by its weight on all the bodies which it surrounds, and also that it is elastic and compressible, he carried a balloon, half-filled with air, to the top of the Puy de Dome, a mountain about 500 toises above Clermont, in Auvergne. It gradually inflated itself as it ascended, and when it reached the summit it was quite full, and swollen as if fresh air had been blown into it; or, what is the same thing, it swelled in proportion as the weight of the column of air which pressed upon it was diminished. When again brought down it became more and more flaccid, and when it reached the bottom it resumed its original condition. In the nine chapters of which the treatise consists, Pascal shows that all the phenomena and effects hitherto ascribed to the horror of a vacuum arise from the weight of the mass of air; and after explaining the variable pressure of the atmosphere in different localities and in its different states, and the rise of water in pumps, he calculates that the whole mass of air round our globe weighs 8,983,889,440,000,000,000 French pounds.—*North-British Review*, No. 2.

It seems probable, from many indications, that the greatest height at which visible clouds *ever exist* does not exceed ten miles; at which height the density of the air is about an eighth part of what it is at the level of the sea.—*Sir John Herschel.*

VARIATIONS OF CLIMATE.

History informs us that many of the countries of Europe which now possess very mild winters, at one time experienced severe cold during this season of the year. The Tiber, at Rome, was often frozen over, and snow at one time lay for forty days in that city.

The Euxine Sea was frozen over every winter during the time of Ovid, and the rivers Rhine and Rhone used to be frozen over so deep that the ice sustained loaded wagons. The waters of the Tiber, Rhine, and Rhone, now flow freely every winter; ice is unknown in Rome, and the waves of the Euxine dash their wintry foam uncrystallised upon the rocks. Some have ascribed these climate changes to agriculture—the cutting down of dense forests, the exposing of the unturned soil to the summer's sun, and the draining of great marshes. We do not believe that such great changes could be produced on the climate of any country by agriculture; and we are certain that no such theory can account for the contrary change of climate—from warm to cold winters— which history tells us has taken place in other countries than those named. Greenland received its name from the emerald herbage which once clothed its valleys and mountains; and its east coast, which is now inaccessible on account of perpetual ice heaped upon its shores, was in the eleventh century the seat of flourishing Scandinavian colonies, all trace of which is now lost. Cold Labrador was named Vinland by the Northmen, who visited it A.D. 1000, and were charmed with its then mild climate. The cause of these changes is an important inquiry.—*Scientific American.*

AVERAGE CLIMATES.

When we consider the numerous and rapid changes which take place in our climate, it is a remarkable fact, that *the mean temperature of a place remains nearly the same.* The winter may be unusually cold, or the summer unusually hot, while the mean temperature has varied even less than a degree. A very warm summer is therefore likely to be accompanied with a cold winter; and in general, if we have any long period of cold weather, we may expect a similar period at a higher temperature. In general, however, in the same locality the relative distribution over summer and winter undergoes comparatively small variations; therefore every point of the globe has an average climate, though it is occasionally disturbed by different atmospheric changes.—*North-British Review*, No. 49.

THE FINEST CLIMATE IN THE WORLD.

Humboldt regards the climate of the Caspian Sea as the most salubrious in the world: here he found the most delicious fruits that he saw during his travels; and such was the purity of the air, that polished steel would not tarnish even by night exposure.

THE PUREST ATMOSPHERES.

The cloudless purity and transparency of the atmosphere, which last for eight months at Santiago, in Chili, are so great, that Lieutenant Gilliss, with the first telescope ever constructed in America, having a diameter of seven inches, was clearly able to recognise the sixth star in the trapezium of Orion. If we are to rely upon the statements of the Rev. Mr. Stoddart, an American missionary, Oroomiah, in Persia, seems to be, in so far as regards the transparency of the atmosphere, the most suitable place in the world for an astronomical observatory. Writing to Sir John Herschel from that country, he mentions that he has been enabled to distinguish with the naked eye the satellites of Jupiter, the crescent of Venus, the rings of Saturn, and the constituent members of several double stars.

SEA-BREEZES AND LAND-BREEZES ILLUSTRATED.

When a fire is kindled on the hearth, we may, if we will observe the motes floating in the room, see that those nearest the chimney are the first to feel the draught and to obey it,—they are drawn into the blaze. The circle of inflowing air is gradually enlarged, until it is scarcely perceived in the remote parts of the room. Now the land is the hearth, the rays of the sun the fire, and the sea, with its cool and calm air, the room; and thus we have at our firesides the sea-breeze in miniature.

When the sun goes down, the fire ceases; then the dry land commences to give off its surplus heat by radiation, so that by nine or ten o'clock it and the air above it are cooled below the sea temperature. The atmosphere on the land thus becomes heavier than that on the sea, and consequently there is a wind seaward, which we call the land-breeze.—*Maury.*

SUPERIOR SALUBRITY OF THE WEST.

All large cities and towns have their best districts in the West;[38] which choice the French *savans*, Pelouze, Pouillet, Boussingault, and Elie de Beaumont, attribute to the law of atmospheric pressure. "When," say they, "the barometric column rises, smoke and pernicious emanations rapidly evaporate in space." On the contrary, smoke and noxious vapours remain in apartments, and on the surface of the soil. Now, of all winds, that which causes the greatest ascension of the barometric column is the east; and that which lowers it most is the west. When the latter blows, it carries with it to the eastern parts of the town all the deleterious gases from the west; and thus the inhabitants of the east have to support their own smoke and miasma, and those brought by western winds. When, on the contrary, the east wind blows, it purifies the air by causing to ascend the pernicious emanations which it cannot drive to the west.

Consequently, the inhabitants of the west receive pure air, from whatever part of the horizon it may arrive; and as the west winds are most prevalent, they are the first to receive the air pure, and as it arrives from the country.

FERTILISATION OF CLOUDS.

As the navigator cruises in the Pacific Ocean among the islands of the trade-wind region, he sees gorgeous piles of cumuli, heaped up in fleecy masses, not only capping the island hills, but often overhanging the lowest islet of the tropics, and even standing above coral patches and hidden reefs; "a cloud by day," to serve as a beacon to the lonely mariner out there at sea, and to warn him of shoals and dangers which no lead nor seaman's eye has ever seen or sounded. These clouds, under favourable circumstances, may be seen gathering above the low coral island, preparing it for vegetation and fruitfulness in a very striking manner. As they are condensed into showers, one fancies that they are a sponge of the most exquisite and delicately elaborated material, and that he can see, as they "drop down their fatness," the invisible but bountiful hand aloft that is pressing and squeezing it out.—*Maury.*

BAROMETRIC MEASUREMENT.

We must not place too implicit a dependence on Barometrical Measurements. Ermann in Siberia, and Ross in the Antarctic Seas, have demonstrated the existence of localities on the earth's surface where a permanent depression of the barometer prevails to the astonishing extent of nearly an inch.

GIGANTIC BAROMETER.

In the Great Exhibition Building of 1851 was a colossal Barometer, the tube and scale reaching from the floor of the gallery nearly to the top of the building, and the rise and fall of the indicating fluid being marked by feet instead of by tenths of inches. The column of mercury, supported by the pressure of the atmosphere, communicated with a perpendicular tube of smaller bore, which contained a coloured fluid much lighter than mercury. When a diminution of atmospheric pressure occurred, the mercury in the large tube descended, and by its fall forced up the coloured fluid in the smaller tube; the fall of the one being indicated in a magnified ratio by the rise in the other.

THE ATMOSPHERE COMPARED TO A STEAM-ENGINE.

In this comparison, by Lieut. Maury, the South Seas themselves, in all their vast intertropical extent, are the boiler for the engine, and the northern hemisphere is its condenser. The mechanical power exerted by the air and the sun in lifting water from the earth, in transporting it from one place to another, and in letting it down again, is inconceivably great. The utilitarian who compares the water-power that the Falls of Niagara would afford if applied to machinery is astonished at the number of figures which are required to express its equivalent in horse-power. Yet what is the horse-power of the Niagara, falling a few steps, in comparison with the horse-power that is required to lift up as high as the clouds and let down again all the water that is discharged into the sea, not only by this river, but by all the other rivers in the world? The calculation has been made by engineers; and according to it, the force of making and lifting vapour from each area of one acre that is included on the surface of the earth, is equal to the power of thirty horses; and for the whole of the earth, it is 800 times greater than all the water-power in Europe.

HOW DOES THE RAIN-MAKING VAPOUR GET FROM THE SOUTHERN INTO THE NORTHERN HEMISPHERE?

This comes with such regularity, that our rivers never go dry, and our springs fail not, because of the exact *compensation* of the grand machine of *the atmosphere*. It is exquisitely and wonderfully counterpoised. Late in the autumn of the north, throughout its winter, and in early spring, the sun is pouring his rays with the greatest intensity down upon the seas of the southern hemisphere; and this powerful engine, which we are contemplating, is pumping up the water there with the greatest activity; at the same time, the mean temperature of the entire southern hemisphere is about 10° higher than the northern. The heat which this heavy evaporation absorbs becomes latent, and with the moisture is carried through the upper regions of the atmosphere until it reaches our climates. Here the vapour is formed into clouds, condensed and precipitated; the heat which held their water in the state of vapour is set free, and becomes sensible heat; and it is that which contributes so much to temper our winter climate. It clouds up in winter, turns warm, and we say we are going to have falling weather: that is because the process of condensation has already commenced, though no rain or snow may have fallen. Thus we feel this southern heat, that has been collected by the rays of the sun by the sea, been bottled away by the winds in the clouds of a southern summer, and set free in the process of condensation in our northern winter.

Thus the South Seas should supply mainly the water for the engine just described, while the northern hemisphere condenses it; we should, therefore, have more rain in the northern hemisphere. The rivers tell us that we have, at least on the land; for the great water-courses of the globe, and half the fresh water in the world, are found on the north side of the equator. This fact is strongly corroborative of this hypothesis. To evaporate water enough annually from the ocean to cover the earth, on the average, five feet deep with rain; to transport it from one zone to another; and to precipitate it in the right places at suitable times and in the proportions due,—is one of the offices of the grand atmospherical machine. This water is evaporated principally from the torrid zone. Supposing it all to come thence, we shall have encircling the earth a belt of ocean 3000 miles in breadth, from which this atmosphere evaporates a layer of water annually sixteen feet in depth. And to hoist up as high as the clouds, and lower down again, all the water, in a lake sixteen feet deep and 3000 miles broad and 24,000 long, is the yearly business of this invisible machinery. What a powerful engine is the atmosphere! and how nicely adjusted must be all the cogs and wheels and springs and *compensations* of this exquisite piece of machinery, that it never wears out nor breaks down, nor fails to do its work at the right time and in the right way!—*Maury.*

THE PHILOSOPHY OF RAIN.

To understand the philosophy of this beautiful and often sublime phenomenon, a few facts derived from observation and a long train of experiments must be remembered.

1. Were the atmosphere every where at all times at a uniform temperature, we should never have rain, or hail, or snow. The water absorbed by it in evaporation from the sea and the earth's surface would descend in an imperceptible vapour, or cease to be absorbed by the air when it was once fully saturated.

2. The absorbing power of the atmosphere, and consequently its capability to retain humidity, is proportionally greater in warm than in cold air.

3. The air near the surface of the earth is warmer than it is in the region of the clouds. The higher we ascend from the earth, the colder do we find the atmosphere. Hence the perpetual snow on very high mountains in the hottest climate.

Now when, from continued evaporation, the air is highly saturated with vapour, though it be invisible and the sky cloudless, if its temperature is suddenly reduced by cold currents descending from above or rushing from a higher to a lower latitude, its capacity to retain moisture is diminished, clouds are formed, and the result is rain. Air condenses as it cools, and, like a sponge filled with water and compressed, pours out the water which its diminished capacity cannot hold. What but Omniscience could have devised such an admirable arrangement for watering the earth?

INORDINATE RAINY CLIMATE.

The climate of the Khasia mountains, which lie north-east from Calcutta, and are separated by the valley of the Burrampooter River from the Himalaya range, is remarkable for the inordinate fall of rain—the greatest, it is said, which has ever been recorded. Mr. Yule, an English gentleman, established that in the single month of August 1841 there fell 264 inches of rain, or 22 feet, of which 12½ feet fell in the space of five consecutive days. This astonishing fact is confirmed by two other

English travellers, who measured 30 inches of rain in twenty-four hours, and during seven months above 500 inches. This great rain-fall is attributed to the abruptness of the mountains which face the Bay of Bengal, and the intervening flat swamps 200 miles in extent. The district of the excessive rain is extremely limited; and but a few degrees farther west, rain is said to be almost unknown, and the winter falls of snow to seldom exceed two inches.

HOW DOES THE NORTH WIND DRIVE AWAY RAIN?

We may liken it to a wet sponge, and the decrease of temperature to the hand that squeezes that sponge. Finally, reaching the cold latitudes, all the moisture that a dew-point of zero, and even far below, can extract, is wrung from it; and this air then commences "to return according to his circuits" as dry atmosphere. And here we can quote Scripture again: "The north wind driveth away rain." This is a meteorological fact of high authority and great importance in the study of the circulation of the atmosphere.—*Maury.*

SIZE OF RAIN-DROPS.

The Drops of Rain vary in their size, perhaps from the 25th to the ¼ of an inch in diameter. In parting from the clouds, they precipitate their descent till the increasing resistance opposed by the air becomes equal to their weight, when they continue to fall with uniform velocity. This velocity is, therefore, in a certain ratio to the diameter of the drops; hence thunder and other showers in which the drops are large pour down faster than a drizzling rain. A drop of the 25th part of an inch, in falling through the air, would, when it had arrived at its uniform velocity, only acquire a celerity of 11½ feet per second; while one of ¼ of an inch would equal a velocity of 33½ feet.—*Leslie.*

RAINLESS DISTRICTS.

In several parts of the world there is no rain at all. In the Old World there are two districts of this kind: the desert of Sahara in Africa, and in Asia part of Arabia, Syria, and Persia; the other district lies between north latitude 30° and 50°, and between 75° and 118° of east longitude, including Thibet, Gobiar Shama, and Mongolia. In the New World the rainless districts are of much less magnitude, occupying two narrow strips on the shores of Peru and Bolivia, and on the coast of Mexico and Guatemala, with a small district between Trinidad and Panama on the coast of Venezuela.

ALL THE RAIN IN THE WORLD.

The Pacific Ocean and the Indian Ocean may be considered as one sheet of water covering an area quite equal in extent to one half of that embraced by the whole surface of the earth; and the total annual fall of rain on the earth's surface is 186,240 cubic imperial miles. Not less than three-fourths of the vapour which makes this rain comes from this waste of waters; but, supposing that only half of this quantity, that is 93,120 cubic miles of rain, falls upon this sea, and that that much at least is taken up from it again as vapour, this would give 255 cubic miles as the quantity of water which is daily lifted up and poured back again into this expanse. It is taken up at one place, and rained down at another; and in this process, therefore, we have agencies for multitudes of partial and conflicting currents, all, in their set strength, apparently as uncertain as the winds.

The better to appreciate the operation of such agencies in producing currents in the sea, imagine a district of 255 square miles to be set apart in the midst of the Pacific Ocean as the scene of operations for one day; then conceive a machine capable of pumping up in the twenty-four hours all the water to the depth of one mile in this district. The machine must not only pump up and bear off this immense quantity of water, but it must discharge it again into the sea on the same day, but at some other place.

All the great rivers of America, Europe, and Asia are lifted up by the atmosphere, and flow in invisible streams back through the air to their sources among the hills; and through channels so regular, certain, and well defined, that the quantity thus conveyed one year with the other is nearly the same: for that is the quantity which we see running down to the ocean through these rivers; and the quantity discharged annually by each river is, as far as we can judge, nearly a constant.—*Maury.*

AN INCH OF RAIN ON THE ATLANTIC.

Lieutenant Maury thus computes the effect of a single Inch of Rain falling upon the Atlantic Ocean. The Atlantic includes an area of twenty-five millions of square miles. Suppose an inch of rain to fall upon only one-fifth of this vast expanse. It would weigh, says our author, three hundred and sixty thousand millions of tons: and the salt which, as water, it held in solution in the sea, and which, when that water was taken up as vapour, was left behind to disturb equilibrium, weighed sixteen millions more of tons, or nearly twice as much as all the ships in the world could carry at a cargo each. It might fall in an hour, or it might fall in a day; but, occupy what time it might in falling, this rain is calculated to exert so much force—which is inconceivably great— in disturbing the equilibrium of the ocean. If all the water discharged by the

Mississippi river during the year were taken up in one mighty measure, and cast into the ocean at one effort, it would not make a greater disturbance in the equilibrium of the sea than would the fall of rain supposed. And yet so gentle are the operations of nature, that movements so vast are unperceived.

THE EQUATORIAL CLOUD-RING.

In crossing the Equatorial Doldrums, the voyager passes a ring of clouds that encircles the earth, and is stretched around our planet to regulate the quantity of precipitation in the rain-belt beneath it; to preserve the due quantum of heat on the face of the earth; to adjust the winds; and send out for distribution to the four corners vapours in proper quantities, to make up to each river-basin, climate, and season, its quota of sunshine, cloud, and moisture. Like the balance-wheel of a well-constructed chronometer, this cloud-ring affords the grand atmospherical machine the most exquisitely arranged *self-compensation*. Nature herself has hung a thermometer under this cloud-belt that is more perfect than any that man can construct, and its indications are not to be mistaken.—*Maury.*

"THE EQUATORIAL DOLDRUMS"

is another of these calm places. Besides being a region of calms and baffling winds, it is a region noted for its rains and clouds, which make it one of the most oppressive and disagreeable places at sea. The emigrant ships from Europe for Australia have to cross it. They are often baffled in it for two or three weeks; then the children and the passengers who are of delicate health suffer most. It is a frightful graveyard on the wayside to that golden land.

BEAUTY OF THE DEW-DROP.

The Dew-drop is familiar to every one from earliest infancy. Resting in luminous beads on the down of leaves, or pendent from the finest blades of grass, or threaded upon the floating lines of the gossamer, its "orient pearl" varies in size from the diameter of a small pea to the most minute atom that can be imagined to exist. Each of these, like the rain-drops, has the properties of reflecting and refracting light; hence, from so many minute prisms, the unfolded rays of the sun are sent up to the eye in colours of brilliancy similar to those of the rainbow. When the sunbeams traverse horizontally a very thickly-bedewed grass-plot, these colours arrange themselves so as to form an iris, or dew-bow; and if we select any one of these drops for observation, and steadily regard it while we gradually change our position, we shall find the prismatic colours follow each other in their regular order.—*Wells.*

FALL OF DEW IN ONE YEAR.

The annual average quantity of Dew deposited in this country is estimated at a depth of about five inches, being about one-seventh of the mean quantity of moisture supposed to be received from the atmosphere all over Great Britain in the year; or about 22,161,337,355 tons, taking the ton at 252 imperial gallons.—*Wells.*

GRADUATED SUPPLY OF DEW TO VEGETATION.

Each of the different grasses draws from the atmosphere during the night a supply of dew to recruit its energies dependent upon its form and peculiar radiating power. Every flower has a power of radiation of its own, subject to changes during the day and night, and the deposition of moisture on it is regulated by the peculiar law which this radiating power obeys; and this power will be influenced by the aspect which the flower presents to the sky, unfolding to the contemplative mind the most beautiful example of creative wisdom.[39]

WARMTH OF SNOW IN ARCTIC LATITUDES.

The first warm Snows of August and September (says Dr. Kane), falling on a thickly-bleached carpet of grasses, heaths, and willows, enshrine the flowery growths which nestle round them in a non-conducting air chamber; and as each successive snow increases the thickness of the cover, we have, before the intense cold of winter sets in, a light cellular bed covered by drift, seven, eight, or ten feet deep, in which the plant retains its vitality. Dr. Kane has proved by experiments that the conducting power of the snow is proportioned to its compression by winds, rains, drifts, and congelation. The drifts that accumulate during nine months of the year are dispersed in well-defined layers of different density. We have first the warm cellular snows of fall, which surround the plant; next the finely-impacted snow-dust of winter; and above these the later humid deposits of spring. In the earlier summer, in the inclined slopes that face the sun, as the upper snow is melted and sinks upon the more compact layer below it is to a great extent arrested, and runs off like rain from a slope of clay. The plant reposes thus in its cellular bed, safe from the rush of waters, and protected from the nightly frosts by the icy roof above it.

IMPURITY OF SNOW.

It is believed that in ascending mountains difficult breathing is sooner felt upon snow than upon rock; and M. Boussingault, in his account of the ascent of Chimborazo, attributes this to the sensible deficiency of oxygen contained in the pores of the snow, which is exhaled when it melts. The fact that the air absorbed by snow is impure, was ascertained by De Saussure, and has been confirmed by Boussingault's experiments.—*Quarterly Review*, No. 202.

SNOW PHENOMENON.

Professor Dove of Berlin relates, in illustration of the formation of clouds of Snow over plains situated at a distance from the cooling summits of mountains, that on one occasion a large company had gathered in a ballroom in Sweden. It was one of those icy starlight nights which in that country are so aptly called "iron nights." The weather was clear and cold, and the ballroom was clear and warm; and the heat was so great, that several ladies fainted. An officer present tried to open a window; but it was frozen fast to the sill. As a last resort, he broke a pane of glass; the cold air rushed in, and it *snowed in the room.* A minute before all was clear; but the warm air of the room had sustained an amount of moisture in a transparent condition which it was not able to maintain when mixed with the colder air from without. The vapour was first condensed, and then frozen.

ABSENCE OF SNOW IN SIBERIA.

There is in Siberia, M. Ermann informs us, an *entire district* in which during the winter the sky is constantly clear, and where a single particle of snow never falls.—*Arago.*

ACCURACY OF THE CHINESE AS OBSERVERS.

The beautiful forms of snow-crystals have long since attracted Chinese observers; for from a remote period there has been met with in their conversation and books an axiomatic expression, to the effect that "snow-flakes are hexagonal," showing the Chinese to be accurate observers of nature.

PROTECTION AGAINST HAIL AND STORMS.

Arago relates, that when, in 1847, two small agricultural districts of Bourgoyne had lost by Hail crops to the value of a million and a half of francs, certain of the proprietors went to consult him on the means of protecting them from like disasters. Resting on the hypothesis of the electric origin of hail, Arago suggested the discharge of the electricity of the clouds by means of balloons communicating by a metallic wire with the soil. This project was not carried out; but Arago persisted in believing in the effectiveness of the method proposed.

Arago, in his *Meteorological Essays,* inquires whether the firing of cannon can dissipate storms. He cites several cases in its favour, and others which seem to oppose it; but he concludes by recommending it to his successors. Whilst Arago was propounding these questions, a person not conversant with science, the poet Méry, was collecting facts supporting the view, which he has published in his *Paris Futur*. His attention was attracted to the firing of cannon to dissipate storms in 1828, whilst an assistant in the "Ecole de Tir" at Vincennes. Having observed that there was never any rain in the morning of the exercise of firing, he waited to examine military records, and found there, as he says, facts which justified the expressions of "Le soleil d'Austerlitz," "Le soleil de juillet," upon the morning of the Revolution of July; and he concluded by proposing to construct around Paris twelve towers of great height, which he calls "tours imbrifuges," each carrying 100 cannons, which should be discharged into the air on the approach of a storm. About this time an incident occurred which in nowise confirmed the truth of M. Méry's theory. The 14th of August was a fine day. On the 15th, the fête of the Empire, the sun shone out, the cannon thundered all day long, fireworks and illuminations were blazing from nine o'clock in the evening. Every thing conspired to verify the hypothesis of M. Méry, and chase away storms for a long time. But towards eleven in the evening a torrent of rain burst upon Paris, in spite of the pretended influence of the discharge of cannon, and gave an occasion for the mobile Gallic mind to turn its attention in other directions.

TERRIFIC HAILSTORM.

Jansen describes, from the log-book of the *Rhijin*, Captain Brandligt, in the South-Indian Ocean (25° south latitude) a Hurricane, accompanied by Hail, by which several of the crew were made blind, others had their faces cut open, and those who were in the rigging had their clothes torn off them. The master of the ship compared the sea "to a hilly landscape in winter covered with snow." Does it not appear as if the "treasures of the hail" were opened, which were "reserved against the time of trouble, against the day of battle and war"?

HOW WATERSPOUTS ARE FORMED IN THE JAVA SEA.

Among the small groups of islands in this sea, in the day and night thunderstorms, the combat of the clouds appears to make them more thirsty than ever. In tunnel form, when they can no longer quench their thirst from the surrounding atmosphere, they descend near the surface of the sea, and appear to lap the water directly up with their black mouths. They are not always accompanied by strong winds; frequently more than one is seen at a time, whereupon the clouds whence they proceed disperse, and the ends of the Waterspouts bending over finally causes them to break in the middle. They seldom last longer than five minutes. As they are going away, the bulbous tube, which is as palpable as that of a thermometer, becomes broader at the base; and little clouds, like steam from the pipe of a locomotive, are continually thrown off from the circumference of the spout, and gradually the water is released, and the cloud whence the spout came again closes its mouth.

COLD IN HUDSON'S BAY.

Mr. R. M. Ballantyne, in his journal of six years' residence in the territories of the Hudson's Bay Company, tells us, that for part of October there is sometimes a little warm, or rather thawy, weather; but after that, until the following April, the thermometer seldom rises to the freezing point. In the depth of winter, the thermometer falls from 30° to 40°, 45°, and even 49° *below zero* of Fahrenheit. This intense cold is not, however, so much felt as one might suppose; for during its continuance the air is perfectly calm. Were the slightest breath of wind to rise when the thermometer stands so low, no man could show his face to it for a moment. Forty degrees below zero, and quite calm, is infinitely preferable to fifteen below, or thereabout, with a strong breeze of wind. Spirit of wine is, of course, the only thing that can be used in the thermometer; as mercury, were it exposed to such cold, would remain frozen nearly half the winter. Spirit never froze in any cold ever experienced at York Factory, unless when very much adulterated with water; and even then the spirit would remain liquid in the centre of the mass. Quicksilver easily freezes in this climate, and it has frequently been run into a bullet-mould, exposed to the cold air till frozen, and in this state rammed down a gun-barrel, and fired through a thick plank. The average cold may be set down at about 15° or 16° below zero, or 48° of frost. The houses at the Bay are built of wood, with double windows and doors. They are heated by large iron stoves, fed with wood; yet so intense is the cold, that when a stove has been in places red-hot, a basin of water in the room has been frozen solid.

PURITY OF WENHAM-LAKE ICE.

Professor Faraday attributes the purity of Wenham-Lake Ice to its being free from air-bubbles and from salts. The presence of the first makes it extremely difficult to succeed in making a lens of English ice which will concentrate the solar rays, and readily fire gunpowder; whereas nothing is easier than to perform this singular feat of igniting a combustible body by aid of a frozen mass if Wenham-Lake ice be employed. The absence of salts conduces greatly to the permanence of the ice; for where water is so frozen that the salts expelled are still contained in air-cavities and cracks, or form thin films between the layers of ice, these entangled salts cause the ice to melt at a lower temperature than 32°, and the liquefied portions give rise to streams and currents within the body of the ice which rapidly carry heat to the interior. The mass then goes on thawing within as well as without, and at temperatures below 32°; whereas pure, compact, Wenham-Lake ice can only thaw at 32°, and only on the outside of the mass.—*Sir Charles Lyell's Second Visit to the United States.*

ARCTIC TEMPERATURES.

Dr. Kane, in his Second Arctic Expedition, found the thermometers beginning to show unexampled temperature: they ranged from 60° to 70° below zero, and upon the taffrail of the brig 65°. The reduced mean of the best spirit-standards gave 67° or 99° below the freezing point of water. At these temperatures chloric ether became solid, and chloroform exhibited a granular pellicle on its surface. Spirit of naphtha froze at 54°, and the oil of turpentine was solid at 63° and 65°.

DR. RAE'S ARCTIC EXPLORATIONS.

The gold medal of the Royal Geographical Society was in 1852 most rightfully awarded to this indefatigable Arctic explorer. His survey of the inlet of Boothia, in 1848, was unique in its kind. In Repulse Bay he maintained his party on deer, principally shot by himself; and spent ten months of an Arctic winter in a hut of stones, with no other fuel than a kind of hay of the *Andromeda tetragona*. Thus he preserved his men to execute surveying journeys of 1000 miles in the spring. Later he travelled 300 miles on snow-shoes. In a spring journey over the ice, with a pound of fat daily for fuel, accompanied by two men only, and trusting solely for shelter to snow-houses, which he taught his men to build, he accomplished 1060 miles in thirty-nine days, or twenty-seven miles per day, including stoppages,—a feat never equalled in Arctic travelling. In the spring journey, and that which followed in the summer in boats, 1700 miles were traversed in eighty days. Dr. Rae's greatest sufferings, he once remarked to Sir George Back, arose from his being obliged to sleep upon his frozen mocassins in order to thaw them for the morning's use.

PHENOMENA OF THE ARCTIC CLIMATE.

Sir John Richardson, in his history of his Expedition to these regions, describes the power of the sun in a cloudless sky to have been so great, that he was glad to take shelter in the water while the crews were engaged on the portages; and he has never felt the direct rays of the sun so oppressive as on some occasions in the high latitudes. Sir John observes:

The rapid evaporation of both snow and ice in the winter and spring, long before the action of the sun has produced the slightest thaw or appearance of moisture, is evident by many facts of daily occurrence. Thus when a shirt, after being washed, is exposed in the open air to a temperature of from 40° to 50° below zero, it is instantly rigidly frozen, and may be broken if violently bent. If agitated when in this condition by a strong wind, it makes a rustling noise like theatrical thunder.

In consequence of the extreme dryness of the atmosphere in winter, most articles of English manufacture brought to Rupert's Land are shrivelled, bent, and broken. The handles of razors and knives, combs, ivory scales, &c., kept in the warm room, are changed in this way. The human body also becomes vividly electric from the dryness of the skin. One cold night I rose from my bed, and was going out to observe the thermometer, with no other clothing than my flannel night-dress, when on my hand approaching the iron latch of the door, a distinct spark was elicited. Friction of the skin at almost all times in winter produced the electric odour.

Even at midwinter we had but three hours and a half of daylight. On December 20th I required a candle to write at the window at ten in the morning. The sun was absent ten days, and its place in the heavens at noon was denoted by rays of light shooting into the sky above the woods.

The moon in the long nights was a most beautiful object, that satellite being constantly above the horizon for nearly a fortnight together. Venus also shone with a brilliancy which is never witnessed in a sky loaded with vapours; and, unless in snowy weather, our nights were always enlivened by the beams of the aurora.

INTENSE HEAT AND COLD OF THE DESERT.

Among crystalline bodies, rock-crystal, or silica, is the best conductor of heat. This fact accounts for the steadiness of temperature in one set district, and the extremes of Heat and Cold presented by day and night on such sandy wastes as the Sahara. The sand, which is for the most part silica, drinks-in the noon-day heat, and loses it by night just as speedily.

The influence of the hot winds from the Sahara has been observed in vessels traversing the Atlantic at a distance of upwards of 1100 geographical miles from the African shores, by the coating of impalpable dust upon the sails.

TRANSPORTING POWER OF WINDS.

The greatest example of their power is the *sand-flood* of Africa, which, moving gradually eastward, has overwhelmed all the land capable of tillage west of the Nile, unless sheltered by high mountains, and threatens ultimately to obliterate the rich plain of Egypt.

EXHILARATION IN ASCENDING MOUNTAINS.

At all elevations of from 6000 to 11,000 feet, and not unfrequently for even 2000 feet more, the pedestrian enjoys a pleasurable feeling, imparted by the consciousness of existence, similar to that which is described as so fascinating by those who have become familiar with the desert-life of the East. The body seems lighter, the nervous power greater, the appetite is increased; and fatigue, though felt for a time, is removed by the shortest repose. Some travellers have described the sensation by the impression that they do not actually press the ground, but that the blade of a knife could be inserted between the sole of the foot and the mountain top.—*Quarterly Review*, No. 202.

TO TELL THE APPROACH OF STORMS.

The proximity of Storms has been ascertained with accuracy by various indications of the electrical state of the atmosphere. Thus Professor Scott, of Sandhurst College, observed in Shetland that drinking-glasses, placed in an inverted position upon a shelf in a cupboard on the ground-floor of Belmont House, occasionally emitted sounds as if they were tapped with a knife, or raised a little and then let fall on the shelf. These sounds preceded wind; and when they occurred, boats and vessels were immediately secured. The strength of the sound is said to be proportioned to the tempest that follows.

REVOLVING STORMS.

By the conjoint labours of Mr. Redfield, Colonel Reid, and Mr. Piddington, on the origin and nature of hurricanes, typhoons, or revolving storms, the following important results have been obtained. Their existence in moderate latitudes on both sides the equator; their absence in the immediate neighbourhood of the equatorial regions; and the fact, that while in the northern latitudes these storms revolve in a direction contrary to the hands of a watch the face of which is placed upwards, in the southern latitudes they rotate in the opposite direction,—are shown to be so many additions to the long chain of evidence by which the rotation of the earth as a physical fact is demonstrated.

IMPETUS OF A STORM.

Captain Sir S. Brown estimates, from experiments made by him at the extremity of the Brighton-Chain Pier in a heavy south-west gale, that the waves impinge on a cylindrical surface one foot high and one foot in diameter with a force equal to eighty pounds, to which must be added that of the wind, which in a violent storm exerts a pressure of forty pounds. He computed the collective impetus of the waves on the lower part of a lighthouse proposed to be built on the Wolf Rock (exposed to the most violent storms of the Atlantic), of the surf on the upper part, and of the wind on the whole, to be equal to 100 tons.

HOW TO MAKE A STORM-GLASS.

This instrument consists of a glass tube, sealed at one end, and furnished with a brass cap at the other end, through which the air is admitted by a very small aperture. Nearly fill the tube with the following solution: camphor, 2½ drams; nitrate of potash, 38 grains; muriate of ammonia, 38 grains; water, 9 drams; rectified spirit, 9 drams. Dissolve with heat. At the ordinary temperature of the atmosphere, plumose crystals are formed. On the approach of stormy weather, these crystals appear compressed into a compact mass at the bottom of the tube; while during fine weather they assume their plumose character, and extend a considerable way up the glass. These results depend upon the condition of the air, but they are not considered to afford any reliable indication of approaching weather.

SPLENDOUR OF THE AURORA BOREALIS.

Humboldt thus beautifully describes this phenomenon:

The intensity of this light is at times so great, that Lowenörn (on June 29, 1786) recognised its coruscation in bright sunshine. Motion renders the phenomenon more visible. Round the point in the vault of heaven which corresponds to the direction of the inclination of the needle the beams unite together to form the so-called corona, the crown of the Northern Light, which encircles the summit of the heavenly canopy with a milder radiance and unflickering emanations of light. It is only in rare instances that a perfect crown or circle is formed; but on its completion, the phenomenon has invariably reached its maximum, and the radiations become less frequent, shorter, and more colourless. The crown, and the luminous arches break up; and the whole vault of heaven becomes covered with irregularly scattered, broad, faint, almost ashy-gray, luminous, immovable patches, which in their turn disappear, leaving nothing but a trace of a dark smoke-like segment on the horizon. There often remains nothing of the whole spectacle but a white delicate cloud with feathery edges, or divided at equal distances into small roundish groups like cirro-cumuli.—*Cosmos*, vol. i.

Among many theories of this phenomenon is that of Lieutenant Hooper, R.N., who has stated to the British Association that he believes "the Aurora Borealis to be no more nor less than the moisture in some shape (whether dew or vapour, liquid or frozen), illuminated by the heavenly bodies, either directly, or reflecting their rays from the frozen masses around the Pole, or even from the immediately proximate snow-clad earth."

VARIETIES OF LIGHTNING.

According to Arago's investigations, the evolution of Lightning is of three kinds: zigzag, and sharply defined at the edges; in sheets of light, illuminating a whole cloud, which seems to open and reveal the light within it; and in the form of fire-balls. The duration of the first two kinds scarcely continues the thousandth part of a second; but the globular lightning moves much more slowly, remaining visible for several seconds.

WHAT IS SHEET-LIGHTNING?

This electric phenomenon is unaccompanied by thunder, or too distant to be heard: when it appears, the whole sky, but particularly the horizon, is suddenly illuminated with a flickering flash. Philosophers differ much as to its cause. Matteucci supposes it to be produced either during evaporation, or evolved (according to Pouillet's theory) in the process of vegetation; or generated by chemical action in the great laboratory of nature, the earth, and accumulated in the lower strata of the air in consequence of the ground being an imperfect conductor.

Arago and Kamtz, however, consider sheet-lightning as *reflections of distant thunderstorms.* Saussure observed sheet-lightning in the direction of Geneva, from the Hospice du Grimsel, on the 10th and 11th of July 1783; while at the same time a terrific thunderstorm raged at Geneva. Howard, from Tottenham, near London, on July 31, 1813, saw sheet-lightning towards the south-east, while the sky was bespangled with stars, not a cloud floating in the air; at the same time a thunderstorm raged at Hastings, and in France from Calais to Dunkirk. Arago supports his opinion, that the phenomenon is *reflected lightning,* by the following illustration: In 1803, when observations were being made for determining the longitude, M. de Zach, on the Brocken, used a few ounces of gunpowder as a signal, the flash of which was visible from the Klenlenberg, sixty leagues off, although these mountains are invisible from each other.

PRODUCTION OF LIGHTNING BY RAIN.

A sudden gust of rain is almost sure to succeed a violent detonation immediately overhead. Mr. Birt, the meteorologist, asks: Is this rain a *cause* or *consequence* of the electric discharge? To this he replies:

In the sudden agglomeration of many minute and feebly electrified globules into one rain-drop, the quantity of electricity is increased in a greater proportion than the surface over which (according to the laws of electric distribution) it is spread. By tension, therefore, it is increased, and may attain the point when it is capable of separating from the *drop* to seek the surface of the *cloud*, or of the newly-formed descending body of rain, which, under such circumstances, may be regarded as a conducting medium. Arrived at this surface, the tension, for the same reason, becomes enormous, and a flash escapes. This theory Mr. Birt has confirmed by observation of rain in thunderstorms.

SERVICE OF LIGHTNING-CONDUCTORS.

Sir David Brewster relates a remarkable instance of a tree in Clandeboye Park, in a thick mass of wood, and *not the tallest of the group*, being struck by lightning, which passed down the trunk into the ground, rending the tree asunder. This shows that an object may be struck by lightning in a locality where there are numerous conducting points more elevated than itself; and at the same time proves that lightning cannot be diverted from its course by lofty isolated conductors, but that the protection of buildings from this species of meteor can only be effected by conductors stretching out in all directions.

Professor Silliman states, that lightning-rods cannot be relied upon unless they reach the earth where it is permanently wet; and that the best security is afforded by carrying the rod, or some good metallic conductor duly connected with it, to the water in the well, or to some other water that never fails. The professor's house, it seems, was struck; but his lightning-rods were not more than two or three inches in the ground, and were therefore virtually of no avail in protecting the building.

ANCIENT LIGHTNING-CONDUCTOR.

Humboldt informs us, that "the most important ancient notice of the relations between lightning and conducting metals is that of Ctesias, in his *Indica*, cap. iv. p. 190. He possessed two iron swords, presents from the king Artaxerxes Mnemon and from his mother Parasytis, which, when planted in the earth, averted clouds, hail, and *strokes of lightning*. He had himself seen the operation, for the king had twice made the experiment before his eyes."—*Cosmos*, vol. ii.

THE TEMPLE OF JERUSALEM PROTECTED FROM LIGHTNING.

We do not learn, either from the Bible or Josephus, that the Temple at Jerusalem was ever struck by Lightning during an interval of more than a thousand years, from the time of Solomon to the year 70; although, from its situation, it was completely exposed to the violent thunderstorms of Palestine.

By a fortuitous circumstance, the Temple was crowned with lightning-conductors similar to those which we now employ, and which we owe to Franklin's discovery. The roof, constructed in what we call the Italian manner, and covered with boards of cedar, having a thick coating of gold, was garnished from end to end with long pointed and gilt iron or steel lances, which, Josephus says, were intended to prevent birds from roosting on the roof and soiling it. The walls were overlaid throughout with wood, thickly gilt. Lastly, there were in the courts of the Temple cisterns, into which the rain from the roof was conducted by *metallic pipes*. We have here both the lightning-rods and a means of conduction so abundant, that Lichtenberg is quite right in saying that many of the present apparatuses are far from offering in their construction so satisfactory a combination of circumstances.—*Abridged from Arago's Meteorological Essays.*

HOW ST. PAUL'S CATHEDRAL IS PROTECTED FROM LIGHTNING.

In March 1769, the Dean and Chapter of St. Paul's addressed a letter to the Royal Society, requesting their opinion as to the best and most effectual method of fixing electrical conductors on the cathedral. A committee was formed for the purpose, and Benjamin Franklin was one of the members; their report was made, and the conductors were fixed as follows:

The seven iron scrolls supporting the ball and cross are connected with other rods (used merely as conductors), which unite them with several large bars, descending obliquely to the stone-work of the lantern, and connected by an iron ring with four other iron bars to the lead covering of the great cupola, a distance of forty-eight feet; thence the communication is continued by the rain-water pipes to the lead-covered roof, and thence by lead water-pipes which pass into the earth; thus completing the entire communication from the cross to the ground, partly through iron, and partly through lead. On the clock-tower a bar of iron connects the pine-apple at the top with the iron staircase, and thence with the lead on the roof of the church. The bell-tower is similarly protected. By these means the metal used in the building is made available as conductors; the metal employed merely for that purpose being exceedingly small in quantity.—*Curiosities of London.*

VARIOUS EFFECTS OF LIGHTNING.

Dr. Hibbert tells us that upon the western coast of Scotland and Ireland, Lightning coöperates with the violence of the storm in shattering solid rocks, and heaping them in piles of enormous fragments, both on dry land and beneath the water.

Euler informs us, in his *Letters to a German Princess*, that he corresponded with a Moravian priest named Divisch, who assured him that he had averted during a whole summer every thunderstorm which threatened his own habitation and the neighbourhood, by means of a machine constructed upon the principles of electricity;

that the machinery sensibly attracted the clouds, and constrained them to descend quietly in a distillation, without any but a very distant thunderclap. Euler assures us that "the fact is undoubted, and confirmed by irresistible proof."

About the year 1811, in the village of Phillipsthal, in Eastern Prussia, an attempt was made to split an immense stone into a multitude of pieces by means of lightning. A bar of iron, in the form of a conductor, was previously fixed to the stone; and the experiment was attended with complete success; for during the very first thunderstorm the lightning burst the stone without displacing it.

The celebrated Duhamel du Monceau says, that lightning, unaccompanied by thunder, wind, or rain, has the property of breaking oat-stalks. The farmers are acquainted with this effect, and say that the lightning breaks down the oats. This is a well-received opinion with the farmers in Devonshire.

Lightning has in some cases the property of reducing solid bodies to ashes, or to pulverisation,—even the human body,—without there being any signs of heat. The effects of lightning on paralysis are very remarkable, in some cases curing, in others causing, that disease.

The returning stroke of lightning is well known to be due to the restoration of the natural electric state, after it has been disturbed by induction.

A THUNDERSTORM SEEN FROM A BALLOON.

Mr. John West, the American aeronaut, in his observations made during his numerous ascents, describes a storm viewed from above the clouds to have the appearance of ebullition. The bulging upper surface of the cloud resembles a vast sea of boiling and upheaving snow; the noise of the falling rain is like that of a waterfall over a precipice; the thunder above the cloud is not loud, and the flashes of lightning appear like streaks of intensely white fire on a surface of white vapour. He thus describes a side view of a storm which he witnessed June 3, 1852, in his balloon excursion from Portsmouth, Ohio:

Although the sun was shining on me, the rain and small hail were rattling on the balloon. A rainbow, or prismatically-coloured arch or horse-shoe, was reflected against the sun; and as the point of observation changed laterally and perpendicularly, the perspective of this golden grotto changed its hues and forms. Above and behind this arch was going on the most terrific thunder; but no zigzag lightning was perceptible, only bright flashes, like explosions of "Roman candles" in fireworks. Occasionally there was a zigzag explosion in the cloud immediately below, the thunder sounding like a *feu-de-joie* of a rifle-corps. Then an orange-coloured wave of light seemed to fall from the upper to the lower cloud; this was "still-lightning." Meanwhile intense electrical action was going on *in the balloon*, such as expansion, tremulous tension, lifting papers ten feet out of the car below the balloon and then dropping them, &c. The close view of this Ohio storm was truly sublime; its rushing noise almost appalling.

Ascending from the earth with a balloon, in the rear of a storm, and mounted up a thousand feet above it, the balloon will soon override the storm, and may descend in advance of it. Mr. West has experienced this several times.

REMARKABLE AERONAUTIC VOYAGE.

Mr. Sadler, the celebrated aeronaut, ascended on one occasion in a balloon from Dublin, and was wafted across the Irish Channel; when, on his approach to the Welsh coast, the balloon descended nearly to the surface of the sea. By this time the sun was set, and the shades of evening began to close in. He threw out nearly all his ballast, and suddenly sprang upward to a great height; and by so doing brought his horizon to *dip* below the sun, producing the whole phenomenon of a western sunrise. Subsequently descending in Wales, he of course witnessed a second sunset on the same evening.—*Sir John Herschel's Outlines of Astronomy.*

Physical Geography of the Sea.40

CLIMATES OF THE SEA.

The fauna and flora of the Sea are as much the creatures of Climate, and are as dependent for their well-being upon temperature, as are the fauna and flora of the dry land. Were it not so, we should find the fish and the algæ, the marine insect and the coral, distributed equally and alike in all parts of the ocean; the polar whale would delight in the torrid zone; and the habitat of the pearl oyster would be also under the iceberg, or in frigid waters colder than the melting ice.

THE CIRCULATION OF THE SEA.

The coral islands, reefs, and beds with which the Pacific Ocean is studded and garnished, were built up of materials which a certain kind of insect quarried from the sea-water. The currents of the sea ministered to this little insect; they were its *hod-carriers*. When fresh supplies of solid matter were wanted for the coral rock upon which the foundations of the Polynesian Islands were laid, these hod-carriers brought them in unfailing streams of sea-water, loaded with food and building-materials for the coralline: the obedient currents thread the widest and the deepest sea. Now we know that its adaptations are suited to all the wants of every one of its inhabitants,—to the wants of the coral insect as well as those of the whale. Hence *we know* that the sea has its system of circulation: for it transports materials for the coral rock from one part of the world to another; its currents receive them from rivers, and hand them over to the little mason for the structure of the most stupendous works of solid masonry that man has ever seen—the coral islands of the sea.

TEMPERATURE OF THE SEA.

Between the hottest hour of the day and the coldest hour of the night there is frequently a change of four degrees in the Temperature of the Sea. Taking one-fifth of the Atlantic Ocean for the scene of operation, and the difference of four degrees to extend only ten feet below the surface, the total and absolute change made in such a mass of sea-water, by altering its temperature two degrees, is equivalent to a change in its volume of 390,000,000 cubic feet.

TRANSPARENCY OF THE OCEAN.

Captain Glynn, U.S.N., has made some interesting observations, ranging over 200° of latitude, in different oceans, in very high latitudes, and near the equator. His apparatus was simple: a common white dinner-plate, slung so as to lie in the water horizontally, and sunk by an iron pot with a line. Numbering the fathoms at which the plate was visible below the surface, Captain Glynn saw it on two occasions, at the maximum, twenty-five fathoms (150 feet) deep; the water was extraordinarily clear, and to lie in the boat and look down was like looking down from the mast-head; and the objects were clearly defined to a great depth.

THE BASIN OF THE ATLANTIC.

In its entire length, the basin of this sea is a long trough, separating the Old World from the New, and extending probably from pole to pole.

This ocean-furrow was scored into the solid crust of our planet by the Almighty hand, that there the waters which "he called seas" might be gathered together so as to "let the dry land appear," and fit the earth for the habitation of man.

From the top of Chimborazo to the bottom of the Atlantic, at the deepest place yet recognised by the plummet in the North Atlantic, the distance in a vertical line is nine miles.

Could the waters of the Atlantic be drawn off, so as to expose to view this great sea-gash, which separates continents, and extends from the Arctic to the Antarctic, it would present a scene the most grand, rugged, and imposing. The very ribs of the solid earth, with the foundations of the sea, would be brought to light; and we should have presented to us at one view, in the empty cradle of the ocean, "a thousand fearful wrecks," with that dreadful array of dead men's skulls, great anchors, heaps of pearls and inestimable stones, which, in the dreamer's eye, lie scattered on the bottom of the sea, making it hideous with sights of ugly death.

GALES OF THE ATLANTIC.

Lieutenant Maury has, in a series of charts of the North and South Atlantic, exhibited, by means of colours, the prevalence of Gales over the more stormy parts of the oceans for each month in the year. One colour shows the region in which there is a gale every six days; another colour every six to ten days; another every ten to fourteen days: and there is a separate chart for each month and each ocean.

SOLITUDE AT SEA.

Between Humboldt's Current of Peru and the great equatorial flow, there is "a desolate region," rarely visited by the whale, either sperm or right. Formerly this part of the ocean was seldom whitened by the sails of a ship, or enlivened by the presence of man. Neither the industrial pursuits of the sea nor the highways of commerce called him into it. Now and then a roving cruiser or an enterprising whalesman passed that way; but to all else it was an unfrequented part of the ocean, and so remained until the gold-fields of Australia and the guano islands of Peru made it a thoroughfare. All vessels bound from Australia to South America now pass through it; and in the journals of some of them it is described as a region almost void of the signs of life in both sea and air. In the South-Pacific Ocean especially, where there is such a wide expanse of water, sea-birds often exhibit a companionship with a vessel, and will follow and keep company with it through storm and calm for weeks together. Even the albatross and Cape pigeon, that delight in the stormy regions of Cape Horn and the inhospitable climates of the Antarctic regions, not unfrequently accompany vessels into the perpetual summer of the tropics. The sea-birds that join the ship as she clears Australia will, it is said, follow her to this region, and then disappear. Even the chirp of the stormy petrel ceases to be heard here, and the sea itself is said to be singularly barren of "moving creatures that have life."

BOTTLES AND CURRENTS AT SEA.

Seafaring people often throw a bottle overboard, with a paper stating the time and place at which it is done. In the absence of other information as to Currents, that afforded by these mute little navigators is of great value. They leave no track behind them, it is true, and their routes cannot be ascertained; but knowing where they are cast, and seeing where they are found, some idea may be formed as to their course. Straight lines may at least be drawn, showing the shortest distance from the beginning to the end of their voyage, with the time elapsed. Admiral Beechey has prepared a chart, representing, in this way, the tracks of more than 100 bottles. From this it appears that the waters from every quarter of the Atlantic tend towards the Gulf of Mexico and its stream. Bottles cast into the sea midway between the Old and the New Worlds, near the coasts of Europe, Africa, and America at the extreme north or farthest south, have been found either in the West Indies, or the British Isles, or within the well-known range of Gulf-Stream waters.

"THE HORSE LATITUDES"

are the belts of calms and light airs which border the polar edge of the north-east trade-winds. They are so called from the circumstance that vessels formerly bound from New England to the West Indies, with a deck-load of horses, were often so delayed in this calm belt of Cancer, that, from the want of water for their animals, they were compelled to throw a portion of them overboard.

"WHITE WATER" AND LUMINOUS ANIMALS AT SEA.

Captain Kingman, of the American clipper-ship *Shooting Star*, in lat. 8° 46′ S., long. 105° 30′ E., describes a patch of *white water*, about twenty-three miles in length, making the whole ocean appear like a plain covered with snow. He filled a 60-gallon tub with the water, and found it to contain small luminous particles seeming to be alive with worms and insects, resembling a grand display of rockets and serpents seen at a great distance in a dark night; some of the serpents appearing to be six inches in length, and very luminous. On being taken up, they emitted light until brought within a few feet of a lamp, when nothing was visible; but by aid of a sextant's magnifier they could be plainly seen—a jelly-like substance, without colour. A specimen two inches long was visible to the naked eye; it was about the size of a large hair, and tapered at the ends. By bringing one end within about one-fourth of an inch of a lighted lamp, the flame was attracted towards it, and burned with a red light; the substance crisped in burning, something like hair, or appeared of a red heat before being consumed. In a glass of the water there were several small round substances (say 1/16th of an inch in diameter) which had the power of expanding and contracting; when expanded, the outer rim appeared like a circular saw, the teeth turned inward.

The scene from the clipper's deck was one of awful grandeur: the sea having turned to phosphorus, and the heavens being hung in blackness, and the stars going out, seemed to indicate that all nature was preparing for that last grand conflagration which we are taught to believe will annihilate this material world.

INVENTION OF THE LOG.

Long before the introduction of the Log, hour-glasses were used to tell the distance in sailing. Columbus, Juan de la Cosa, Sebastian Cabot, and Vasco de Gama, were not acquainted with the Log and its mode of application; and they estimated the ship's speed merely by the eye, while they found the distance they had made by the running-down of the sand in the *ampotellas*, or hour-glasses. The Log for the measurement of the distance traversed is stated by writers on navigation not to have been invented until the end of the sixteenth or the beginning of the seventeenth century (see *Encyclopædia Britannica*, 7th edition, 1842). The precise date is not known; but it is certain that Pigafetta, the companion of Magellan, speaks, in 1521, of the Log as a well-known means of finding the course passed over. Navarete places the use of the log-line in English ships in 1577.

LIFE OF THE SEA-DEEPS.

The ocean teems with life, we know. Of the four elements of the old philosophers,—fire, earth, air, and water,—perhaps the sea most of all abounds with living creatures. The space occupied on the surface of our planet by the different families of animals and their remains is inversely as the size of the individual; the smaller the animal, generally speaking, the greater the space occupied by his remains. Take the elephant and his remains, and a microscopic animal and his, and compare them; the contrast as to space occupied is as striking as that of the coral reef or island with the dimensions of the whale. The graveyard that would hold the corallines, is larger than the graveyard that would hold the elephants.

DEPTHS OF OCEAN AND AIR UNKNOWN.

At some few places under the tropics, no bottom has been found with soundings of 26,000 feet, or more than four miles; whilst in the air, if, according to Wollaston, we may assume that it has a limit from which waves of sound may be reverberated, the phenomenon of twilight would incline us to assume a height at least nine times as great. The aerial ocean rests partly on the solid earth, whose mountain-chains and elevated plateaus rise like green wooded shoals, and partly on the sea, whose surface forms a moving base, on which rest the lower, denser, and more saturated strata of air.—*Humboldt's Cosmos*, vol. i.

The old Alexandrian mathematicians, on the testimony of Plutarch, believed the depth of the sea to depend on the height of the mountains. Mr. W. Darling has propounded to the British Association the theory, that as the sea covers three times the area of the land, so it is reasonable to suppose that the depth of the ocean, and that for a large portion, is three times as great as the height of the highest mountain. Recent soundings show depths in the sea much greater than any elevations on the surface of the earth; for a line has been veered to the extent of seven miles.—*Dr. Scoresby.*

GREATEST ASCERTAINED DEPTH OF THE SEA.

In the dynamical theory of the tides, the ratio of the effects of the sun and moon depends, not only on the masses, distances, and periodic times of the two luminaries, but also on the Depth of the Sea; and this, accordingly, may be computed when the other quantities are known. In this manner Professor Haughton has deduced, from the solar and lunar coefficients of the diurnal tide, a mean depth of 5•12 miles; a result which accords in a remarkable manner with that inferred from the ratio of the semi-diurnal co-efficients as obtained by Laplace from the Brest observations. Professor Hennessey states, that from what is now known regarding the depth of the ocean, the continents would appear as plateaus elevated above the oceanic depressions to an amount which, although small compared to the earth's radius, would be considerable when compared to its outswelling at the equator and its flattening towards the poles; and the surface thus presented would be the true surface of the earth.

The greatest depths at which the bottom of the sea has been reached with the plummet are in the North-Atlantic Ocean; and the places where it has been fathomed (by the United-States deep-sea sounding apparatus) do not show it to be deeper than 25,000 feet = 4 miles, 1293 yards, 1 foot. The deepest place in this ocean is probably between the parallels of 35° and 40° north latitude, and immediately to the southward of the Grand Banks of Newfoundland.

It appears that, with one exception, the bottom of the North-Atlantic Ocean, as far as examined, from the depth of about sixty fathoms to that of more than two miles (2000 fathoms), is literally nothing but a mass of microscopic shells. Not one of the animalcules from these shells has been found living in the surface-waters, nor in shallow water along the shore. Hence arises the question, Do they live on the bottom, at the immense depths where the shells are found; or are they borne by submarine currents from their real habitat?

RELATIVE LEVELS OF THE RED SEA AND MEDITERRANEAN.

The French engineers, at the beginning of the present century, came to the conclusion that the Red Sea was about thirty feet above the Mediterranean: but the observations of Mr. Robert Stephenson, the English engineer, at Suez; of M. Negretti, the Austrian, at Tineh, near the ancient Pelusium; and the levellings of Messrs. Talabat, Bourdaloue, and their assistants between the two seas;—have proved that the low-water mark of ordinary tides at Suez and Tineh is very nearly on the same levels, the difference being that at Suez it is rather more than one inch lower.—*Leonard Horner; Proceedings of the Royal Society*, 1855.

THE DEPTH OF THE MEDITERRANEAN.

Soundings made in the Mediterranean suffice to indicate depths equal to the average height of the mountains girding round this great basin; and, if one particular experiment may be credited, reaching even to 15,000 feet—an equivalent to the elevation of the highest Alps. This sounding was made about ninety miles east of Malta. Between Cyprus and Egypt, 6000 feet of line had been let down without reaching the bottom. Other deep soundings have been made in other places with similar results. In the lines of sea between Egypt and the Archipelago, it is stated that one sounding made by the *Tartarus* between Alexandria and Rhodes reached bottom at the depth of 9900 feet; another, between Alexandria and Candia, gave a depth of 300 feet beyond this. These single soundings, indeed, whether of ocean or sea, are always open to the certainty that greater as well as lesser depths must exist, to which no line has ever been sunk; a case coming under that general law of probabilities so largely applicable in every part of physics. In the Mediterranean especially, which has so many aspects of a sunken basin, there may be abysses of depth here and there which no plummet is ever destined to reach.—*Edinburgh Review.*

COLOUR OF THE RED SEA.

M. Ehrenberg, while navigating the Red Sea, observed that the red colour of its waters was owing to enormous quantities of a new animal, which has received the name of *oscillatoria rubescens,* and which seems to be the same with what Haller has described as a *purple conferva* swimming in water; yet Dr. Bonar, in his work entitled *The Desert of Sinai,* records:

Blue I have called the sea; yet not strictly so, save in the far distance. It is neither a *red* nor a *blue* sea, but emphatically green,—yes, green, of the most brilliant kind I ever saw. This is produced by the immense tracts of shallow water, with yellow sand beneath, which always gives this green to the sea, even in the absence of verdure on the shore or sea-weeds beneath. The *blue* of the sky and the *yellow* of the sands meeting and intermingling in the water, form the *green* of the sea; the water being the medium in which the mixing or fusing of the colours takes place.

WHAT IS SEA-MILK?

The phenomena with this name and that of "Squid" are occasioned by the presence of phosphorescent animalcules. They are especially produced in the intertropical seas, and they appear to be chiefly abundant in the Gulf of Guinea and in the Arabian Gulf. In the latter, the phenomenon was known to the ancients more than a century before the Christian era, as may be seen from a curious passage from the geography of

Agatharcides: "Along this country (the coast of Arabia) the sea has a white aspect like a river: the cause of this phenomenon is a subject of astonishment to us." M. Quatrefages has discovered that the *Noctilucæ* which produce this phenomenon do not always give out clear and brilliant sparks, but that under certain circumstances this light is replaced by a steady clearness, which gives in these animalcules a white colour. The waters in which they have been observed do not change their place to any sensible degree.

THE BOTTOM OF THE SEA A BURIAL-PLACE.

Among the minute shells which have been fished up from the great telegraphic plateau at the bottom of the sea between Newfoundland and Ireland, the microscope has failed to detect a single particle of sand or gravel; and the inference is, that there, if any where, the waters of the sea are at rest. There is not motion enough there to abrade these very delicate organisms, nor current enough to sweep them about and mix them up with a grain of the finest sand, nor the smallest particle of gravel from the loose beds of *débris* that here and there strew the bottom of the sea. The animalculæ probably do not live or die there. They would have had no light there; and, if they lived there, their frail textures would be subjected in their growth to a pressure upon them of a column of water 12,000 feet high, equal to the weight of 400 atmospheres. They probably live and sport near the surface, where they can feel the genial influence of both light and heat, and are buried in the lichen caves below after death.

It is now suggested, that henceforward we should view the surface of the sea as a nursery teeming with nascent organisms, and its depths as the cemetery for families of living creatures that outnumber the sands on the sea-shore for multitude.

Where there is a nursery, hard by there will be found also a graveyard,—such is the condition of the animal world. But it never occurred to us before to consider the surface of the sea as one wide nursery, its every ripple as a cradle, and its bottom one vast burial-place.—*Lieut. Maury.*

WHY IS THE SEA SALT?

It has been replied, In order to preserve it in a state of purity; which is, however, untenable, mainly from the fact that organic impurities in a vast body of moving water, whether fresh or salt, become rapidly lost, so as apparently to have called forth a special agency to arrest the total organised matter in its final oscillation between the organic and inorganic worlds. Thus countless hosts of microscopic creatures swarm in most waters, their principal function being, as Professor Owen surmises, to feed upon and thus restore to the living chain the almost unorganised matter of various zones. These creatures preying upon one another, and being preyed upon by others in their turn, the circulation of organic matter is kept up. If we do not adopt this view, we must at least look upon the Infusoria and Foraminifera as scavenger agents to prevent an undue accumulation of decaying matter; and thus the salt condition of the sea is not a necessity.

Nor is the amount of saline matter in the sea sufficient to arrest decomposition. That the sea is salt to render it of greater density, and by lowering its freezing point to preserve it from congelation to within a shorter distance of the poles, though admissible, scarcely meets the entire solution of the question. The freezing point of sea-water, for instance, is only $3\frac{1}{2}°$ F. lower than that of fresh water; hence, with the present distribution of land and sea—and still less, probably, with that which obtained in former geological epochs—no very important effects would have resulted had the ocean been fresh instead of salt.

Now Professor Chapman, of Toronto, suggests that the salt condition of the sea is mainly intended to regulate evaporation, and to prevent an undue excess of that phenomenon; saturated solutions evaporating more slowly than weak ones, and these latter more slowly again than pure water.

Here, then, we have a self-adjusting phenomenon and admirable contrivance in the balance of forces. If from any temporary cause there be an unusual amount of saline matter in the sea, evaporation goes on the more and more slowly; and, on the other hand, if this proportion be reduced by the addition of fresh water in undue excess, the evaporating power is the more and more increased—thus aiding time, in either instance, to restore the balance. The perfect system of oceanic circulation may be ascribed, in a great degree at least, if not wholly, to the effect produced by the salts of the sea upon the mobility and circulation of its waters.

Now this is an office which the sea performs in the economy of the universe by virtue of its saltness, and which it could not perform were its waters altogether fresh. And thus philosophers have a clue placed in their hands which will probably guide to one of the many hidden reasons that are embraced in the true answer to the question, "*Why is the sea salt?*"

HOW TO ASCERTAIN THE SALTNESS OF THE SEA.

Dry a towel in the sun, weigh it carefully, and note its weight. Then dip it into sea-water, wring it sufficiently to prevent its dripping, and weigh it again; the increase of the weight being that of the water imbibed by the cloth. It should then be thoroughly dried, and once more weighed; and the excess of this weight above the original weight of the cloth shows the quantity of the salt retained by it; then, by comparing the weight of this salt with that of the sea-water imbibed by the cloth, we shall find what proportion of salt was contained in the water.

ALL THE SALT IN THE SEA.

The amount of common Salt in all the oceans is estimated by Schafhäutl at 3,051,342 cubic geographical miles. This would be about five times more than the mass of the Alps, and only one-third less than that of the Himalaya. The sulphate of soda equals 633,644•36 cubic miles, or is equal to the mass of the Alps; the chloride of magnesium, 441,811•80 cubic miles; the lime salts, 109,339•44 cubic miles. The above supposes the mean depth to be but 300 metres, as estimated by Humboldt. Admitting, with Laplace, that the mean depth is 1000 metres, which is more probable, the mass of marine salt will be more than double the mass of the Himalaya.—*Silliman's Journal*, No. 16.

Taking the average depth of the ocean at two miles, and its average saltness at 3½ per cent, it appears that there is salt enough in the sea to cover to the thickness of one mile an area of 7,000,000 of square miles. Admit a transfer of such a quantity of matter from an average of half a mile above to one mile below the sea-level, and astronomers will show by calculation that it would alter the length of the day.

These 7,000,000 of cubic miles of crystal salt have not made the sea any fuller.

PROPERTIES OF SEA-WATER.

The solid constituents of sea-water amount to about 3½ per cent of its weight, or nearly half an ounce to the pound. Its saltness is caused as follows: Rivers which are constantly flowing into the ocean contain salts varying from 10 to 50, and even 100, grains per gallon. They are chiefly common salt, sulphate and carbonate of lime, magnesia,41 soda, potash, and iron; and these are found to constitute the distinguishing characteristics of sea-water. The water which evaporates from the sea is nearly pure, containing but very minute traces of salts. Falling as rain upon the land, it washes the soil, percolates through the rocky layers, and becomes charged with saline substances, which are borne seaward by the returning currents. The ocean, therefore, is the great depository of every thing that water can dissolve and carry down from the surface of the continents; and as there is no channel for their escape, they consequently accumulate (*Youmans' Chemistry*). They would constantly accumulate, as this very shrewd author remarks, were it not for the shells and insects of the sea and other agents.

SCENERY AND LIFE OF THE ARCTIC REGIONS.

The late Dr. Scoresby, from personal observations made in the course of twenty-one voyages to the Arctic Regions, thus describes these striking characteristics:

The coast scenes of Greenland are generally of an abrupt character, the mountains frequently rising in triangular profile; so much so, that it is sometimes not possible to effect their ascent. One of the most notable characteristics of the Arctic lands is the deception to which travellers are liable in regard to distances. The occasion of this is the quantity of light reflected from the snow, contrasted with the dark colour of the rocks. Several persons of considerable experience have been deceived in this way, imagining, for example, that they were close to the shore when in fact they were more than twenty miles off. The trees of these lands are not more than three inches above ground.

Many of the icebergs are five miles in extent, and some are to be seen running along the shore measuring as much as thirteen miles. Dr. Scoresby has seen a cliff of ice supported on those floating masses 402 feet in height. There is no place in the world where animal life is to be found in greater profusion than in Greenland, Spitzbergen, Baffin's Bay, and other portions of the Arctic regions. This is to be accounted for by the abundance and richness of the food supplied by the sea. The number of birds is especially remarkable. On one occasion, no less than a million of little hawks came in sight of Dr. Scoresby's ship within a single hour.

The various phenomena of the Greenland sea are very interesting. The different colours of the sea-water—olive or bottle-green, reddish-brown, and mustard—have, by the aid of the microscope, been found to be owing to animalculæ of these various colours: in a single drop of mustard-coloured water have been counted 26,450 animals. Another remarkable characteristic of the Greenland sea-water is its warm temperature—one, two, and three degrees above the freezing-point even in the cold season. This Dr. Scoresby accounts for by supposing the flow in that direction of warm currents from the south. The polar fields of ice are to be found from eight or nine to thirty or forty feet in thickness. By fastening a hook twelve or twenty inches in these masses of ice, a ship could ride out in safety the heaviest gales.

ICEBERG OF THE POLAR SEAS.

The ice of this berg, although opaque and vascular, is true glacier ice, having the fracture, lustre, and other external characters of a nearly homogeneous growth. The iceberg is true ice, and is always dreaded by ships. Indeed, though modified by climate, and especially by the alternation of day and night, the polar glacier must be regarded as strictly atmospheric in its increments, and not essentially differing from the glacier of the Alps. The general appearance of a berg may be compared to frosted silver; but when its fractures are very extensive, the exposed faces have a very brilliant lustre. Nothing can be more exquisite than a fresh, cleanly fractured berg surface: it reminds one of the recent cleavage of sulphate of strontian—a resemblance more striking from the slightly lazulitic tinge of each.—*U. S. Grinnel Expedition in Search of Sir J. Franklin.*

IMMENSITY OF POLAR ICE.

The quantity of solid matter that is drifted out of the Polar Seas through one opening—Davis's Straits—alone, and during a part of the year only, covers to the depth of seven feet an area of 300,000 square miles, and weighs not less than 18,000,000,000 tons. The quantity of water required to float and drive out this solid matter is probably many times greater than this. A quantity of water equal in weight to these two masses has to go in. The basin to receive these inflowing waters, i. e. the unexplored basin about the North Pole, includes an area of 1,500,000 square miles; and as the outflowing ice and water are at the surface, the return current must be submarine.

These two currents, therefore, it may be perceived, keep in motion between the temperate and polar regions of the earth a volume of water, in comparison with which the mighty Mississippi in its greatest floods sinks down to a mere rill.—*Maury.*

OPEN SEA AT THE POLE.

The following fact is striking: In 1662–3, Mr. Oldenburg, Secretary to the Royal Society, was ordered to register a paper entitled "Several Inquiries concerning Greenland, answered by Mr. Gray, who had visited those parts." The nineteenth query was, "How near any one hath been known to approach the Pole. *Answer.* I once met upon the coast of Greenland a Hollander, that swore he had been but half a degree from the Pole, showing me his journal, which was also attested by his mate; where *they had seen no ice or land, but all water.*" Boyle mentions a similar account, which he received from an old Greenland master, on April 5, 1765.

RIVER-WATER ON THE OCEAN.

Captain Sabine found discoloured water, supposed to be that of the Amazon, 300 miles distant in the ocean from the embouchure of that river. It was about 126 feet deep. Its specific gravity was $= 1 \cdot 0204$, and the specific gravity of the sea-water $= 1 \cdot 0262$. This appears to be the greatest distance from land at which river-water has been detected on the surface of the ocean. It was estimated to be moving at the rate of three miles an hour, and had been turned aside by an ocean-current. "It is not a little curious to reflect," says Sir Henry de la Beche, "that the agitation and resistance of its particles should be sufficient to keep finely comminuted solid matter mechanically suspended, so that it would not be disposed freely to part with it except at its junction with the sea-water over which it flows, and where, from friction, it is sufficiently retarded."

THE THAMES AND ITS SALT-WATER BED.

The Thames below Woolwich, in place of flowing upon a solid bottom, really flows upon the liquid bottom formed by the water of the sea. At the flow of the tide, the fresh water is raised, as it were, in a single mass by the salt water which flows in, and which ascends the bed of the river, while the fresh water continues to flow towards the sea.—*Mr. Stevenson, in Jameson's Journal.*

FRESH SPRINGS IN THE MIDDLE OF THE OCEAN.

On the southern coast of the island of Cuba, at a few miles from land, Springs of Fresh Water gush from the bed of the Ocean, probably under the influence of hydrostatic pressure, and rise through the midst of the salt water. They issue forth with such force that boats are cautious in approaching this locality, which has an ill repute on account of the high cross sea thus caused. Trading vessels sometimes visit these springs to take in a supply of fresh water, which is thus obtained in the open sea. The greater the depth from which the water is taken, the fresher it is found to be.

"THE BLACK WATERS."

In the upper portion of the basin of the Orinoco and its tributaries, Nature has several times repeated the enigmatical phenomenon of the so-called "Black Waters." The Atabapo, whose banks are adorned with Carolinias and arborescent Melastomas, is a river of a coffee-brown colour. In the shade of the palm-groves this colour seems about to pass into ink-black. When placed in transparent vessels, the water appears of a golden yellow. The image of the Southern Constellation is reflected with wonderful clearness in these black streams. When their waters flow gently, they afford to the observer, when taking astronomical observations with reflecting instruments, a most excellent artificial horizon. These waters probably owe their peculiar colour to a solution of carburetted hydrogen, to the luxuriance of the tropical vegetation, and to the quantity of plants and herbs on the ground over which they flow.—*Humboldt's Aspects of Nature*, vol. i.

GREAT CATARACT IN INDIA.

Where the river Shirhawti, between Bombay and Cape Comorin, falls into the Gulf of Arabia, it is about one-fourth of a mile in width, and in the rainy season some thirty feet in depth. This immense body of water rushes down a rocky slope 300 feet, at an angle of 45°, at the bottom of which it makes a perpendicular plunge of 850 feet into a black and dismal abyss, with a noise like the loudest thunder. The whole descent is therefore 1150 feet, or several times that of Niagara; but the volume of water in the latter is somewhat larger than in the former.

CAUSE OF WAVES.

The friction of the wind combines with the tide in agitating the surface of the ocean, and, according to the theory of undulations, each produces its effect independently of the other. Wind, however, not only raises waves, but causes a transfer of superficial water also. Attraction between the particles of air and water, as well as the pressure of the atmosphere, brings its lower stratum into adhesive contact with the surface of the sea. If the motion of the wind be parallel to the surface, there will still be friction, but the water will be smooth as a mirror; but if it be inclined, in however small a degree, a ripple will appear. The friction raises a minute wave, whose elevation protects the water beyond it from the wind, which consequently impinges on the surface at a small angle: thus each impulse, combining with the other, produces an undulation which continually advances.—*Mrs. Somerville's Physical Geography.*

RATE AT WHICH WAVES TRAVEL.

Professor Bache states, as one of the effects of an earthquake at Simoda, on the island of Niphon, in Japan, that the harbour was first emptied of water, and then came in an enormous wave, which again receded and left the harbour dry. This occurred several times. The United-States self-acting tide-gauge at San Francisco, which records the rise of the tide upon cylinders turned by clocks, showed that at San Francisco, 4800 miles from the scene of the earthquake, the first wave arrived twelve hours and sixteen minutes after it had receded from the harbour of Simoda. It had travelled across the broad bosom of the Pacific Ocean at the rate of six miles and a half a minute, and arrived on the shores of California: the first wave being seven-tenths of a foot in height, and lasting for about half an hour, followed by seven lesser waves, at intervals of half an hour each.

The velocity with which a wave travels depends on the depth of the ocean. The latest calculations for the Pacific Ocean give a depth of from 14,000 to 18,000 fathoms. It is remarkable how the estimates of the ocean's depth have grown less. Laplace assumed it at ten miles, Whewell at 3·5, while the above estimate brings it down to two miles.

Mr. Findlay states, that the dynamic force exerted by Sea-Waves is greatest at the crest of the wave before it breaks; and its power in raising itself is measured by various facts. At Wasburg, in Norway, in 1820, it rose 400 feet; and on the coast of Cornwall, in 1843, 300 feet. The author shows that waves have sometimes raised a column of water equivalent to a pressure of from three to five tons the square foot. He also proves that the velocity of the waves depends on their length, and that waves of from 300 to 400 feet in length from crest to crest travel from twenty to twenty-seven and a half miles an hour. Waves travel great distances, and are often raised by distant hurricanes, having been felt simultaneously at St. Helena and Ascension, though 600 miles apart; and it is probable that ground-swells often originate at the Cape of Good Hope, 3000 miles distant. Dr. Scoresby found the travelling rate of the Atlantic waves to be 32•67 English statute miles per hour.

In the winter of 1856, a heavy ground-swell, brought on by five hours' gale, scoured away in fourteen hours 3,900,000 tons of pebbles from the coast near Dover; but in three days, without any shift of wind, upwards of 3,000,000 tons were thrown back again. These figures are to a certain extent conjectural; but the quantities have been derived from careful measurement of the profile of the beach.

OCEAN-HIGHWAYS: HOW SEA-ROUTES HAVE BEEN SHORTENED.

When one looks seaward from the shore, and sees a ship disappear in the horizon as she gains an offing on a voyage to India, or the Antipodes perhaps, the common idea is that she is bound over a trackless waste; and the chances of another ship sailing with the same destination the next day, or the next week, coming up and speaking with her on the "pathless ocean," would to most minds seem slender indeed. Yet the truth is, the winds and the currents are now becoming so well understood, that the navigator, like the backwoodsman in the wilderness, is enabled literally to "blaze his way" across the ocean; not, indeed, upon trees, as in the wilderness, but upon the wings of the wind. The results of scientific inquiry have so taught him how to use these invisible couriers, that they, with the calm belts of the air, serve as sign-boards to indicate to him the turnings and forks and crossings by the way.

Let a ship sail from New York to California, and the next week let a faster one follow; they will cross each other's path many times, and are almost sure to see each other by the way, as in the voyage of two fine clipper-ships from New York to California. On the ninth day after the *Archer* had sailed, the *Flying Cloud* put to sea. Both ships were running against time, but without reference to each other. The *Archer*, with wind and current charts in hand, went blazing her way across the calms of Cancer, and along the new route down through the north-east trades to the equator; the *Cloud* followed, crossing the equator upon the trail of Thomas of the *Archer*. Off Cape Horn she came up with him, spoke him, and handed him the latest New York dates. The *Flying Cloud* finally ranged ahead, made her adieus, and disappeared among the clouds that lowered upon the western horizon, being destined to reach her port a week or more in advance of her Cape Horn consort. Though sighting no land from the time of their separation until they gained the offing of San Francisco,—some six or eight thousand miles off,—the tracks of the two vessels were so nearly the same, that being projected upon the chart, they appear almost as one.

This is the great course of the ocean: it is 15,000 miles in length. Some of the most glorious trials of speed and of prowess that the world ever witnessed among ships that "walk the waters" have taken place over it. Here the modern clipper-ship—the noblest work that has ever come from the hands of man—has been sent, guided by the lights of science, to contend with the elements, to outstrip steam, and astonish the world.—*Maury.*

ERROR UPON ERROR.

The great inducement to Mr. Babbage, some years since, to attempt the construction of a machine by which astronomical tables could be calculated and even printed by mechanical means, and with entire accuracy, was the errors in the requisite tables. Nineteen such errors, in point of fact, were discovered in an edition of Taylor's *Logarithms* printed in 1796; some of which might have led to the most dangerous results in calculating a ship's place. These nineteen errors (of which one only was an error of the press) were pointed out in the *Nautical Almanac* for 1832. In one of these *errata*, the seat of the error was stated to be in cosine of 14° 18′ 3″. Subsequent examination showed that there was an error of one second in this correction, and accordingly, in the *Nautical Almanac* of the next year a new correction was necessary. But in making the new correction of one second, a new error was committed of ten degrees, making it still necessary, in some future edition of the *Nautical Almanac*, to insert an *erratum* in an *erratum* of the *errata* in Taylor's *Logarithms.—Edinburgh Review*, vol. 59.

Phenomena of Heat.

THE LENGTH OF THE DAY AND THE HEAT OF THE EARTH.

As we may judge of the uniformity of temperature from the unaltered time of vibration of a pendulum, so we may also learn from the unaltered rotatory velocity of the earth the amount of stability in the mean temperature of our globe. This is the result of one of the most brilliant applications of the knowledge we had long possessed of the movement of the heavens to the thermic condition of our planet. The rotatory velocity of the earth depends on its volume; and since, by the gradual cooling of the mass by radiation, the axis of rotation would become shorter, the rotatory velocity would necessarily increase, and the length of the day diminish with a decrease of the temperature. From the comparison of the secular inequalities in the motions of the moon with the eclipses observed in former ages, it follows that, since the time of Hipparchus,—that is, for full 2000 years,—the length of the day has certainly not diminished by the hundredth part of a second. The decrease of the mean heat of the globe during a period of 2000 years has not therefore, taking the extremest limits, diminished as much as 1/306th of a degree of Fahrenheit.[42]—*Humboldt's Cosmos*, vol. i.

NICE MEASUREMENT OF HEAT.

A delicate thermometer, placed on the ground, will be affected by the passage of a single cloud across a clear sky; and if a succession of clouds pass over, with intervals of clear sky between them, such an instrument has been observed to fluctuate accordingly, rising with each passing mass of vapour, and falling again when the radiation becomes unrestrained.

EXPENDITURE OF HEAT BY THE SUN.

Sir John Herschel estimates the total Expenditure of Heat by the Sun in a given time, by supposing a cylinder of ice 45 miles in diameter to be continually darted into the sun *with the velocity of light*, and that the water produced by its fusion were continually carried off: the heat now given off constantly by radiation would then be wholly expended in its liquefaction, on the one hand, so as to leave no radiant surplus; while, on the other, the actual temperature at its surface would undergo no diminution.

The great mystery, however, is to conceive how so enormous a conflagration (if such it be) can be kept up. Every discovery in chemical science here leaves us completely at a loss, or rather seems to remove further the prospect of probable explanation. If conjecture might be hazarded, we should look rather to the known possibility of an indefinite generation of heat by friction, or to its excitement by the electric discharge, than to any combustion of ponderable fuel, whether solid or gaseous, for the origin of the solar radiation.—*Outlines.* 43

DISTINCTIONS OF HEAT.

Among the curious laws of modern science are those which regulate the transmission of radiant heat through transparent bodies. The heat of our fires is intercepted and detained by screens of glass, and, being so detained, warms them; while solar heat passes freely through and produces no such effect. "The more recent researches of Delaroche," says Sir John Herschel, "however, have shown that this detention is complete only when the temperature of the source of heat is low; but that as the temperature gets higher a portion of the heat radiated acquires a power of penetrating glass, and that the quantity which does so bears continually a larger and larger proportion to the whole, as the heat of the radiant body is more intense. This discovery is very important, as it establishes a community of nature between solar and terrestrial heat; while at the same time it leads us to regard the actual temperature of the sun as far exceeding that of any earthly flame."

LATENT HEAT.

This extraordinary principle exists in all bodies, and may be pressed out of them. The blacksmith hammers a nail until it becomes red hot, and from it he lights the match with which he kindles the fire of his forge. The iron has by this process become more dense, and percussion will not again produce incandescence until the bar has been exposed in fire to a red heat, when it absorbs heat, the particles are restored to their former state, and we can again by hammering develop both heat and light.—*R. Hunt, F.R.S.*

HEAT AND EVAPORATION.

In a communication made to the French Academy, M. Daubrée calculates that the Evaporation of the Water on the surface of the globe employs a quantity of heat about equal to one-third of what is received from the sun; or, in other words, equal to the melting of a bed of ice nearly thirty-five feet in thickness if spread over the globe.

HEAT AND MECHANICAL POWER.

It has been found that Heat and Mechanical Power are mutually convertible; and that the relation between them is definite, 772 foot-pounds of motive power being equivalent to a unit of heat, that is, to the amount of heat requisite to raise a pound of water through one degree of Fahrenheit.

HEAT OF MINES.

One cause of the great Heat of many of our deep Mines, which appears to have been entirely lost sight of, is the chemical action going on upon large masses of pyritic matter in their vicinity. The heat, which is so oppressive in the United Mines in Cornwall that the miners work nearly naked, and bathe in water at 80° to cool themselves, is without doubt due to the decomposition of immense quantities of the sulphurets of iron and copper known to be in this condition at a short distance from these mineral works.—*R. Hunt, F.R.S.*

VIBRATION OF HEATED METALS.

Mr. Arthur Trevelyan discovered accidentally that a bar of iron, when heated and placed with one end on a solid block of lead, in cooling vibrates considerably, and produces sounds similar to those of an Æolian harp. The same effect is produced by bars of copper, zinc, brass, and bell-metal, when heated and placed on blocks of lead, tin, or pewter. The bars were four inches long, one inch and a half wide, and three-eighths of an inch thick.

The conditions essential to these experiments are, That two different metals must be employed—the one soft and possessed of moderate conducting powers, viz. lead or tin, the other hard; and it matters not whether soft metal be employed for the bar or block, provided the soft metal be cold and the hard metal heated.

That the surface of the block shall be uneven, for when rendered quite smooth the vibration does not take place; but the bar cannot be too smooth.

That no matter be interposed, else it will prevent vibration, with the exception of a burnish of gold leaf, the thickness of which cannot amount to the two-hundred-thousandth part of an inch.—*Transactions of the Royal Society of Edinburgh.*

EXPANSION OF SPIRITS.

Spirits expand and become lighter by means of heat in a greater proportion than water, wherefore they are heaviest in winter. A cubic inch of brandy has been found by many experiments to weigh ten grains more in winter than in summer, the difference being between four drams thirty-two grains and four drams forty-two grains. Liquor-merchants take advantage of this circumstance, and make their purchases in winter rather than in summer, because they get in reality rather a larger quantity in the same bulk, buying by measure.—*Notes in Various Sciences.*

HEAT PASSING THROUGH GLASS.

The following experiment is by Mr. Fox Talbot: Heat a poker bright-red hot, and having opened a window, apply the poker quickly very near to the outside of a pane, and the hand to the inside; a strong heat will be felt at the instant, which will cease as soon as the poker is withdrawn, and may be again renewed and made to cease as quickly as before. Now it is well known, that if a piece of glass is so much warmed as to convey the impression of heat to the hand, it will retain some part of that heat for a minute or more; but in this experiment the heat will vanish in a moment: it will not, therefore, be the heated pane of glass that we shall feel, but heat which has come through the glass in a free or radiant state.

HEAT FROM GAS-LIGHTING.

In the winter of 1835, Mr. W. H. White ascertained the temperature in the City to be 3° higher than three miles south of London Bridge; and *after the gas had been lighted in the City* four or five hours the temperature increased full 3°, thus making 6° difference in the three miles.

HEAT BY FRICTION.

Friction as a source of Heat is well known: we rub our hands to warm them, and we grease the axles of carriage-wheels to prevent their setting fire to the wood. Count Rumford has established the extraordinary fact, that an unlimited supply of heat may be derived from friction by the same materials: he made great quantities of water boil by causing a blunt borer to rub against a mass of metal immersed in the water. Savages light their fires by rubbing two pieces of wood: the *modus operandi*, as practised by the Kaffirs of South Africa, is thus described by Captain Drayton:

Two dry sticks, one being of hard and the other of soft wood, were the materials used. The soft stick was laid on the ground, and held firmly down by one Kaffir, whilst another employed himself in scooping out a little hole in the centre of it with the point of his assagy: into this little hollow the end of the hard wood was placed, and held vertically. These two men sat face to face, one taking the vertical stick between the palms of his hands, and making it twist about very quickly, while the other Kaffir held the lower stick firmly in its place; the friction caused by the end of one piece of wood revolving upon the other soon made the two pieces smoke. When the Kaffir who twisted became tired, the respective duties were exchanged. These operations having continued about a couple of minutes, sparks began to appear, and when they became numerous, were gathered into some dry grass, which was then swung round at arm's length until a blaze was established; and a roaring fire was gladdening the hearts of the Kaffirs with the anticipation of a glorious feast in about ten minutes from the time that the operation was first commenced.

HEAT BY FRICTION FROM ICE.

When Sir Humphry Davy was studying medicine at Penzance, one of his constant associates was Mr. Tom Harvey, a druggist in the above town. They constantly experimented together; and one severe winter's day, after a discussion on the nature of heat, the young philosophers were induced to go to Larigan river, where Davy succeeded in developing heat by *rubbing two pieces of ice together* so as to melt each other;[44] an experiment which he repeated with much *éclat* many years after, in the zenith of his celebrity, at the Royal Institution. The pieces of ice for this experiment are fastened to the ends of two sticks, and rubbed together in air below the temperature of 32°: this Davy readily accomplished on the day of severe cold at the Larigan river; but when the experiment was repeated at the Royal Institution, it was in the vacuum of an air-pump, when the temperature of the apparatus and of the surrounding air was below 32°. It was remarked, that when the surface of the rubbing pieces was rough, only half as much heat was evolved as when it was smooth. When the pressure of the rubbing piece was increased four times, the proportion of heat evolved was increased sevenfold.

WARMING WITH ICE.

In common language, any thing is understood to be cooled or warmed when the temperature thereof is made higher or lower, whatever may have been the temperature when the change was commenced. Thus it is said that melted iron is *cooled* down to a sub-red heat, or mercury is cooled from the freezing point to zero, or far below. By the same rule, solid mercury, say 50° below zero, may, in any climate or temperature of the atmosphere, be immediately warmed and melted by being imbedded in a cake of ice.—*Scientific American.*

REPULSION BY HEAT.

If water is poured upon an iron sieve, the wires of which are made red-hot, it will not run through; but on cooling, it will pass through rapidly. M. Boutigny, pursuing this curious inquiry, has proved that the moisture upon the skin is sufficient to protect it from disorganisation if the arm is plunged into baths of melted metal. The resistance of the surfaces is so great that little elevation of temperature is experienced. Professor Plücker has stated, that by washing the arm with ether previously to plunging it into melted metal, the sensation produced while in the molten mass is that of freezing coldness.—*R. Hunt, F.R.S.*

PROTECTION FROM INTENSE HEAT.

The singular power which the body possesses of resisting great heats, and of breathing air of high temperatures, has at various times excited popular wonder. In the last century some curious experiments were made on this subject. Sir Joseph Banks, Dr. Solander, and Sir Charles Blagden, entered a room in which the air had a temperature of 198° Fahr., and remained ten minutes. Subsequently they entered the room separately, when Dr. Solander found the heat 210°, and Sir Joseph 211°, whilst their bodies preserved their natural degree of heat. Whenever they breathed upon a thermometer, it sank several degrees; every inspiration gave coolness to their nostrils, and their breath cooled their fingers when it reached them. Sir Charles Blagden entered an apartment when the heat was 1° or 2° above 260°, and remained eight minutes, mostly on the coolest spot, where the heat was above 240°. Though very hot, Sir Charles felt no pain: during seven minutes his breathing was good; but he then felt an oppression in his lungs, and his pulse was 144, double its ordinary quickness. To prove the heat of the room, eggs and a beefsteak were placed upon a tin frame near the thermometer, when in twenty minutes the eggs were roasted hard, and in forty-seven minutes the steak was dressed dry; and when the air was put in motion by a pair of bellows upon another steak, part of it was well done in thirteen minutes. It is remarkable, that in these experiments the same person who experienced no inconvenience from air heated to 211°, could just bear rectified spirits of wine at 130°, cooling oil at 129°, cooling water at 123°, and cooling quicksilver at 117°.

Sir Francis Chantrey, the sculptor, however, exposed himself to a temperature still higher than any yet mentioned, as described by Sir David Brewster:

The furnace which he employs for drying his moulds is about fourteen feet long, twelve feet high, and twelve feet broad. When it is raised to its highest temperature, with the doors closed, the thermometer stands at 350°, and the iron floor is red-hot. The workmen often enter it at a temperature of 340°, walking over the iron floor with wooden clogs, which are of course charred on the surface. On one occasion, Mr. Chantrey, accompanied by five or six of his friends, entered the furnace; and after remaining two minutes they brought out a thermometer which stood at 320°. Some of the party experienced sharp pains in the tips of their ears and in the septum of the nose, while others felt a pain in their eyes.—*Natural Magic*, 1833.

In some cases the clothing worn by the experimenters conducts away the heat. Thus, in 1828, a Spaniard entered a heated oven, at the New Tivoli, near Paris; he sang a song while a fowl was roasted by his side, he then ate the fowl and drank a bottle of wine, and on coming out his pulse beat 176°, and the thermometer was at 110° Reaumur. He then stretched himself upon a plank in the oven surrounded by lighted candles, when the mouth of the oven was closed; he remained there five minutes, and on being taken out, all the candles were extinguished and melted, and the Spaniard's pulse beat 200°. Now much of the surprise ceases when it is added that he wore wide woollen pantaloons, a loose mantle of wool, and a great quilted cap; the several materials of this clothing being bad conductors of heat.

In 1829 M. Chabert, the "Fire-King," exhibited similar feats at the Argyll Rooms in Regent Street. He first swallowed forty grains of phosphorus, then two spoonfuls of oil at 330°, and next held his head over the fumes of sulphuric acid. He had previously provided himself with an antidote for the poison of the phosphorus. Dressed in a loose woollen coat, he then entered a heated oven, and in five minutes cooked two steaks; he then came out of the oven, when the thermometer stood at 380°. Upon another occasion, at White Conduit House, some of his feats were detected.

The scientific secret is as follows: Muscular tissue is an extremely bad conductor; and to this in a great measure the constancy of the temperature of the human body in various zones is to be attributed. To this fact also Sir Charles Blagden and Chantrey owed their safety in exposing their bodies to a high temperature; from the almost impervious character of the tissues of the body, the irritation produced was confined to the surface.

Magnetism and Electricity.

MAGNETIC HYPOTHESES.

As an instance of the obstacles which erroneous hypotheses throw in the way of scientific discovery, Professor Faraday adduces the unsuccessful attempts that had been made in England to educe Magnetism from Electricity until Oersted showed the simple way. Faraday relates, that when he came to the Royal Institution as an assistant in the laboratory, he saw Davy, Wollaston, and Young trying, by every way that suggested itself to them, to produce magnetic effects from an electric current; but having their minds diverted from the true course by their existing hypotheses, it did not occur to them to try the effect of holding a wire through which an electric current was passing over a suspended magnetic needle. Had they done so, as Oersted afterwards did, the immediate deflection of the needle would have proved the magnetic property of an electric current. Faraday has shown that the magnetism of a steel bar is caused by the accumulated action of all the particles of which it is composed: this he proves by first magnetising a small steel bar, and then breaking it successively into smaller and smaller pieces, each one of which possesses a separate pole; and the same operation may be continued until the particles become so small as not to be distinguishable without a microscope.

We quote the above from a late Number of the *Philosophical Magazine*, wherein also we find the following noble tribute to the genius and public and private worth of Faraday:

The public never can know and appreciate the national value of such a man as Faraday. He does not work to please the public, nor to win its guineas; and the said public, if asked its opinion as to the practical value of his researches, can see no possible practical issue there. The public does not know that we need prophets more than mechanics in science,—inspired men, who, by patient self-denial and the exercise of the high intellectual gifts of the Creator, bring us intelligence of His doings in Nature. To them their pursuits are good in themselves. Their chief reward is the delight of being admitted into communion with Nature, the pleasure of tracing out and proclaiming her laws, wholly forgetful whether those laws will ever augment our banker's account or improve our knowledge of cookery. *Such men, though not honoured by the title of "practical," are they which make practical men possible.* They bring us the tamed forces of Nature, and leave it to others to contrive the machinery to which they may be yoked. If we are rightly informed, it was Faradaic electricity which shot the glad tidings of the fall of Sebastopol from Balaklava to Varna. Had this man converted his talent to commercial purposes, as so many do, we should not like to set a limit to his professional income. The quality of his services cannot be expressed by pounds; but that brave body, which for forty years has been the instrument of that great soul, is a fit object for a nation's care, as the achievements of the man are, or will one day be, the object of a nation's pride and gratitude.

THE CHINESE AND THE MAGNETIC NEEDLE.

More than a thousand years before our era, a people living in the extremest eastern portions of Asia had magnetic carriages, on which the movable arm of the figure of a man continually pointed to the south, as a guide by which to find the way across the boundless grass-plains of Tartary; nay, even in the third century of our era, therefore at least 700 years before the use of the mariner's compass in European seas, Chinese vessels navigated the Indian Ocean under the direction of Magnetic Needles pointing to the south.

Now the Western nations, the Greeks and the Romans, knew that magnetism could be communicated to iron, and *that that metal* would retain it for a length of time. The great discovery of the terrestrial directive force depended, therefore, alone on this—that no one in the West had happened to observe an elongated fragment of magnetic iron-stone, or a magnetic iron rod, floating by the aid of a piece of wood in water, or suspended in the air by a thread, in such a position as to admit of free motion.—*Humboldt's Cosmos*, vol. i.

KIRCHER'S "MAGNETISM."

More than two centuries since, Athanasius Kircher published his strange book on Magnetism, in which he anticipated the supposed virtue of magnetic traction in the curative art, and advocated the magnetism of the sun and moon, of the divining-rod, and showed his firm belief in animal magnetism. "In speaking of the vegetable world," says Mr. Hunt, "and the remarkable processes by which the leaf, the flower, and the fruit are produced, this sage brings forward the fact of the diamagnetic (repelled by the magnet) character of the plant which was in 1852 rediscovered; and he refers the motions of the sunflower, the closing of the convolvulus, and the directions of the spiral formed by the twining plants, to this particular influence."[45] Nor were Kircher's anticipations random guesses, but the result of deductions from experiment and observation; and the universality of magnetism is now almost recognised by philosophers.

MINUTE MEASUREMENT OF TIME.

By observing the magnet in the highly-convenient and delicate manner introduced by Gauss and Weber, which consists in attaching a mirror to the magnet and determining the constant factor necessary to convert the differences of oscillation into differences of time, Professor Helmholtz has been able, with comparatively simple apparatus, to make accurate determinations up to the 1/10000th part of a second.

POWER OF A MAGNET.

The Power of a Magnet is estimated by the weight its poles are able to carry. Each pole singly is able to support a smaller weight than when they both act together by means of a keeper, for which reason horse-shoe magnets are superior to bar magnets of similar dimensions and character. It has further been ascertained that small magnets have a much greater relative force than large ones.

When magnetism is excited in a piece of steel in the ordinary mode, by friction with a magnet, it would seem that its inductive power is able to overcome the coercive power of the steel only to a certain depth below the surface; hence we see why small pieces of steel, especially if not very hard, are able to carry greater relative weights than large magnets. Sir Isaac Newton wore in a ring a magnet weighing only 3 grains, which would lift 760 grains, *i. e.* 250 times its own weight.

Bar-magnets are seldom found capable of carrying more than their own weight; but horse-shoe magnets of similar steel will bear considerably more. Small ones of from half an ounce to 1 ounce in weight will carry from 30 to 40 times their own weight; while such as weigh from 1 to 2 lbs. will rarely carry more than from 10 to 15 times their weight. The writer found a 1 lb. horse-shoe magnet that he impregnated by means of the feeder able to bear 26½ times its own weight; and Fischer, having adopted the like mode of magnetising the steel, which he also carefully heated, has made magnets of from 1 to 3 lbs. weight that would carry 30 times, and others of from 4 to 6 lbs. weight that would carry 20 times, their own weight.—*Professor Peschel.*

HOW ARTIFICIAL MAGNETS ARE MADE.

In 1750, Mr. Canton, F.R.S., "one of the most successful experimenters in the golden age of electricity,"[46] communicated to the Royal Society his "Method of making Artificial Magnets without the use of natural ones." This he effected by using a poker and tongs to communicate magnetism to steel bars. He derived his first hint from observing them one evening, as he was sitting by the fire, to be nearly in the same direction with the earth as the dipping needle. He thence concluded that they must, from their position and the frequent blows they receive, have acquired some magnetic virtue, which on trial he found to be the case; and therefore he employed them to impregnate his bars, instead of having recourse to the natural loadstone. Upon the reading of the above paper, Canton exhibited to the Royal Society his experiments, for which the Copley Medal was awarded to him in 1751.

Canton had, as early as 1747, turned his attention, with complete success, to the production of powerful artificial magnets, principally in consequence of the expense of procuring those made by Dr. Gowan Knight, who kept his process secret. Canton for several years abstained from communicating his method even to his most intimate friends, lest it might be injurious to Dr. Knight, who procured considerable pecuniary advantages by touching needles for the mariner's compass.

At length Dr. Knight's method of making artificial magnets was communicated to the world by Mr. Wilson, in a paper published in the 69th volume of the *Philosophical Transactions*. He provided himself with a large quantity of clean iron-filings, which he put into a capacious tub about half full of clear water; he then agitated the tub to and fro for several hours, until the filings were reduced by attrition to an almost impalpable powder. This powder was then dried, and formed into paste by admixture with linseed-oil. The paste was then moulded into convenient shapes, which were exposed to a moderate heat until they had attained a sufficient degree of hardness.

After allowing them to remain for some time in this state, Dr. Knight gave them their magnetic virtue in any direction he pleased, by placing them between the extreme ends of his large magazine of artificial magnets for a second or more, as he saw occasion. By this method the virtue they acquired was such, that when any one of these pieces was held between two of his best ten-guinea bars, with its poles purposely inverted, it immediately of itself turned about to recover its natural direction, which the force of those very powerful bars was not sufficient to counteract.

Dr. Knight's powerful battery of magnets above mentioned is in the possession of the Royal Society, having been presented by Dr. John Fothergill in 1776.

POWER OF THE SUN'S RAYS IN INCREASING THE STRENGTH OF MAGNETS.

Professor Barlocci found that an armed natural loadstone, which would carry 1½ Roman pounds, had its power nearly *doubled* by twenty-four hours' exposure to the strong light of the sun. M. Zantedeschi found that an artificial horse-shoe loadstone, which carried 13½ oz., carried 3½ more by three days' exposure, and at last arrived to 31 oz. by continuing it in the sun's light. He found that while the strength increased in oxidated magnets, it diminished in those which were not oxidated, the diminution becoming insensible when the loadstone was highly polished. He now concentrated the solar rays upon the loadstone by means of a lens; and he found that, both in oxidated and polished magnets, they *acquire* strength when their *north* pole is exposed to the sun's rays, and *lose* strength when the *south* pole is exposed.—*Sir David Brewster.*

COLOUR OF A BODY AND ITS MAGNETIC PROPERTIES.

Solar rays bleach dead vegetable matter with rapidity, while in living parts of plants their action is frequently to strengthen the colour. Their power is perhaps best seen on the sides of peaches, apples, &c., which, exposed to a midsummer's sun, become highly coloured. In the open winter of 1850, Mr. Adie, of Liverpool, found in a wallflower plant proof of a like effect: in the dark months there was a slow succession of one or two flowers, of uniform pale yellow hue; in March streaks of a darker colour appeared on the flowers, and continued to slowly increase till in April they were variegated brown and yellow, of rich strong colours. On the supposition that these changes are referable to magnetic properties, may hereafter be explained Mrs. Somerville's experiments on steel needles exposed to the sun's rays under envelopes of silk of various colours; the magnetisation of steel needles has failed in the coloured rays of the spectrum, but Mr. Adie considers that under dyed silk the effect will hinge on the chemical change wrought in the silk and its dye by the solar rays.

THE ONION AND MAGNETISM.

A popular notion has long been current, more especially on the shores of the Mediterranean, that if a magnetic rod be rubbed with an onion, or brought in contact with the emanations of the plant, the directive force will be diminished, while a compass thus treated will mislead the steersman. It is difficult to conceive what could have given rise to so singular a popular error.47—*Humboldt's Cosmos*, vol. v.

DECLINATION OF THE NEEDLE—THE EARTH A MAGNET.

The Inclination or Dip of the Needle was first recorded by Robert Norman, in a scarce book published in 1576 entitled *The New Attractive; containing a short Discourse of the Magnet or Loadstone, &c.*

Columbus has not only the merit of being the first to discover *a line without magnetic variation*, but also of having first excited a taste for the study of terrestrial magnetism in Europe, by means of his observations on the progressive increase of western declination in receding from that line.

The first chart showing the variation of the compass,48 or the declination of the needle, based on the idea of employing curves drawn through points of equal declination, is due to Halley, who is justly entitled the father and founder of terrestrial magnetism. And it is curious to find that in No. 195 of the *Philosophical Transactions*, in 1683, Halley had previously expressed his belief that he has put it past doubt that the globe of the earth is one great magnet, having four magnetical poles or points of attraction, near each pole of the equator two; and that in those parts of the world which lie near adjacent to any one of those magnetical poles, the needle is chiefly governed thereby, the nearest pole being always predominant over the more remote.

"To Halley" (says Sir John Herschel) "we owe the first appreciation of the real complexity of the subject of magnetism. It is wonderful indeed, and a striking proof of the penetration and sagacity of this extraordinary man, that with his means of information he should have been able to draw such conclusions, and to take so large and comprehensive a view of the subject as he appears to have done."

And, in our time, "the earth is a great magnet," says Faraday: "its power, according to Gauss, being equal to that which would be conferred if every cubic yard of it contained six one-pound magnets; the sum of the force is therefore equal to 8,464,000,000,000,000,000,000 such magnets."

THE AURORA BOREALIS.

Halley, upon his return from his voyage to verify his theory of the variation of the compass, in 1700, hazarded the conjecture that the Aurora Borealis is a magnetic phenomenon. And Faraday's brilliant discovery of the evolution of light by magnetism has raised Halley's hypothesis, enounced in 1714, to the rank of an experimental certainty.

EFFECT OF LIGHT ON THE MAGNET.

In 1854, Sir John Ross stated to the British Association, in proof of the effect of every description of light on the magnet, that during his last voyage in the *Felix*, when frozen in about one hundred miles north of the magnetic pole, he concentrated the rays of the full moon on the magnetic needle, when he found it was five degrees attracted by it.

MAGNETO-ELECTRICITY.

In 1820, the Copley Medal was adjudicated to M. Oersted of Copenhagen, "when," says Dr. Whewell, "the philosopher announced that the conducting-wire of a voltaic circuit acts upon a magnetic needle; and thus recalled into activity that endeavour to connect magnetism with electricity which, though apparently on many accounts so hopeful, had hitherto been attended with no success. Oersted found that the needle has a tendency to place itself at *right angles* to the wire; a kind of action altogether different from any which had been suspected."

ELECTRO-MAGNETS OF THE HORSE-SHOE FORM

were discovered by Sturgeon in 1825. Of two Magnets made by a process devised by M. Elias, and manufactured by M. Logemeur at Haerlem, one, a single horse-shoe magnet weighing about 1 lb., lifts 28½ lbs.; the other, a triple horse-shoe magnet of about 10 lbs. weight, is capable of lifting about 150 lbs. Similar magnets are made by the same person capable of supporting 5 cwt. In the process of making them, a helix of copper and a galvanic battery are used. The smaller magnet has twice the power expressed by Haecker's formula for the best artificial steel magnet.

Subsequently Henry and Ten Eyk, in America, constructed some electro-magnets on a large scale. One horse-shoe magnet made by them, weighing 60 lbs., would support more than 2000 lbs.

In September 1858, there were constructed for the Atlantic-telegraph cable at Valentia two permanent magnets, from which the electric induction is obtained: each is composed of 30 horse-shoe magnets, 2½ feet long and from 4 to 5 inches broad; the induction coils attached to these each contain six miles of wire, and a shock from them, if passed through the human body, would be sufficient to destroy life.

ROTATION-MAGNETISM.

The unexpected discovery of Rotation-Magnetism by Arago, in 1825, has shown practically that every kind of matter is susceptible of magnetism; and the recent investigations of Faraday on diamagnetic substances have, under special conditions of meridian or equatorial direction, and of solid, fluid, or gaseous inactive conditions of the bodies, confirmed this important result.

INFLUENCE OF PENDULUMS ON EACH OTHER.

About a century since it became known, that when two clocks are in action upon the same shelf, they will disturb each other: that the pendulum of the one will stop that of the other; and that the pendulum that was stopped will after a while resume its vibrations, and in its turn stop that of the other clock. When two clocks are placed near one another in cases very slightly fixed, or when they stand on the boards of a floor, they will affect a little each other's pendulum. Mr. Ellicote observed that two clocks resting against the same rail, which agreed to a second for several days, varied one minute thirty-six seconds in twenty-four hours when separated. The slower, having a longer pendulum, set the other in motion in 16-1/3 minutes, and stopped itself in 36-2/3 minutes.

WEIGHT OF THE EARTH ASCERTAINED BY THE PENDULUM.

By a series of comparisons with Pendulums placed at the surface and the interior of the Earth, the Astronomer-Royal has ascertained the variation of gravity in descending to the bottom of a deep mine, as the Harton coal-pit, near South Shields. By calculations from these experiments, he has found the mean density of the earth to be 6•566, the specific gravity of water being represented by unity. In other words, it has been ascertained by these experiments that if the earth's mass possessed every where its average density, it would weigh, bulk for bulk, 6•566 times as much as water. It is curious to note the different values of the earth's mean density which have been obtained by different methods. The Schehallien experiment indicated a mean density equal to about 4½; the Cavendish apparatus, repeated by Baily and Reich, about 5½; and Professor Airy's pendulum experiment furnishes a value amounting to about 6½.

The immediate result of the computations of the Astronomer-Royal is: supposing a clock adjusted to go true time at the top of the mine, it would gain 2¼ seconds per day at the bottom. Or it may be stated thus: that gravity is greater at the bottom of a mine than at the top by 1/19190th part.—*Letter to James Mather, Esq., South Shields.* See also *Professor Airy's Lecture*, 1854.

ORIGIN OF TERRESTRIAL MAGNETISM.

The earliest view of Terrestrial Magnetism supposed the existence of a magnet at the earth's centre. As this does not accord with the observations on declination, inclination, and intensity, Tobias Meyer gave this fictitious magnet an eccentric position, placing it one-seventh part of the earth's radius from the centre. Hansteen imagined that there were two such magnets, different in position and intensity. Ampère set aside these unsatisfactory hypotheses by the view, derived from his discovery, that the earth itself is an electro-magnet, magnetised by an electric current circulating about it from east to west perpendicularly to the plane of the magnetic meridian, to which the same currents give direction as well as magnetise the ores of iron: the currents being thermo-electric currents, excited by the action of the sun's heat successively on the different parts of the earth's surface as it revolves towards the east.

William Gilbert,[49] who wrote an able work on magnetic and electric forces in the year 1600, regarded terrestrial magnetism and electricity as two emanations of a single fundamental source pervading all matter, and he therefore treated of both at once. According to Gilbert's idea, the earth itself is a magnet; whilst he considered that the inflections of the lines of equal declination and inclination depend upon the distribution of mass, the configuration of continents, or the form and extent of the deep intervening oceanic basins.

Till within the last eighty years, it appears to have been the received opinion that the intensity of terrestrial magnetism was the same at all parts of the earth's surface. In the instructions drawn up by the French Academy for the expedition under La Pérouse, the first intimation is given of a contrary opinion. It is recommended that the time of vibration of a dipping-needle should be observed at stations widely remote, as a test of the equality or difference of the magnetic intensity; suggesting also that such observations should particularly be made at those parts of the earth where the dip was greatest and where it was least. The experiments, whatever their results may have been, which, in compliance with this recommendation, were made in the expedition of La Pérouse, perished in its general catastrophe; but the instructions survived.

In 1811, Hansteen took up the subject, and in 1819 published his celebrated work, clearly demonstrating the fluctuations which this element has undergone during the last two centuries; confirming in great detail the position of Halley, that "the whole magnetic system is in motion, that the moving force is very great as extending its effects from pole to pole, and that its motion is not *per saltum*, but a gradual and regular motion."

THE NORTH AND SOUTH MAGNETIC POLES.

The knowledge of the geographical position of both Magnetic Poles is due to the scientific energy of the same navigator, Sir James Ross. His observations of the Northern Magnetic Pole were made during the second expedition of his uncle, Sir John Ross (1829–1833); and of the Southern during the Antarctic expedition under his own command (1839–1843). The Northern Magnetic Pole, in 70° 5' lat., 96° 43' W. long., is 5° of latitude farther from the ordinary pole of the earth than the Southern Magnetic Pole, 75° 35' lat., 154° 10' E. long.; whilst it is also situated farther west from Greenwich than the Northern Magnetic Pole. The latter belongs to the great island of Boothia Felix, which is situated very near the American continent, and is a portion of the district which Captain Parry had previously named North Somerset. It is not far distant from the western coast of Boothia Felix, near the promontory of Adelaide, which extends into King William's Sound and Victoria Strait.

The Southern Magnetic Pole has been directly reached in the same manner as the Northern Pole. On 17th February 1841, the *Erebus* penetrated as far as 76° 12' S. lat., and 164° E. long. As the inclination was here only 88° 40', it was assumed that the Southern Magnetic Pole was about 160 nautical miles distant. Many accurate observations of declination, determining the intersection of the magnetic meridian, render it very probable that the South Magnetic Pole is situated in the interior of the great Antarctic region of South Victoria Land, west of the Prince Albert mountains, which approach the South Pole and are connected with the active volcano of Erebus, which is 12,400 feet in height.—*Humboldt's Cosmos*, vol. v.

MAGNETIC STORMS.

The mysterious course of the magnetic needle is equally affected by time and space, by the sun's course, and by changes of place on the earth's surface. Between the tropics the hour of the day may be known by the direction of the needle as well as by the oscillations of the barometer. It is affected instantly, but transiently, by the northern light.

When the uniform horary motion of the needle is disturbed by a magnetic storm, the perturbation manifests itself *simultaneously*, in the strictest sense of the word, over hundreds and thousands of miles of sea and land, or propagates itself by degrees in short intervals every where over the earth's surface.

Among numerous examples of perturbations occurring simultaneously and extending over wide portions of the earth's surface, one of the most remarkable is that of September 25th, 1841, which was observed at Toronto in Canada, at the Cape of Good Hope, at Prague, and partially in Van Diemen's Land. Sabine adds, "The English Sunday, on which it is deemed sinful, after midnight on Saturday, to register an observation, and to follow out the great phenomena of creation in their perfect development, interrupted the observation in Van Diemen's Land, where, in consequence of the difference of the longitude, the magnetic storm fell on Sunday."

It is but justice to add, that to the direct instrumentality of the British Association we are indebted for this system of observation, which would not have been possible without some such machinery for concerted action. It being known that the magnetic needle is subject to oscillations, the nature, the periods, and the laws of which were unascertained, under the direction of a committee of the Association *magnetic observatories* were established in various places for investigating these strange disturbances. As might have been anticipated, regularly recurring perturbations were noted, depending on the hour of the day and the season of the year. Magnetic storms were observed to sweep simultaneously over the whole face of the earth, and these too have now been ascertained to follow certain periodic laws.

But the most startling result of the combined magnetic observations is the discovery of marked perturbations recurring at intervals of ten years; a period which seemed to have no analogy to any thing in the universe, but which M. Schwabe has found to correspond with the variation of the spots on the sun, both attaining their maximum and minimum developments at the same time. Here, for the present, the discovery stops; but that which is now an unexplained coincidence may hereafter supply the key to the nature and source of Terrestrial Magnetism: or, as Dr. Lloyd observes, this system of magnetic observation has gone beyond our globe, and opened a new range for inquiry, by showing us that this wondrous agent has power in other parts of the solar system.

FAMILIAR GALVANIC EFFECTS.

By means of the galvanic agency a variety of surprising effects have been produced. Gunpowder, cotton, and other inflammable substances have been set on fire; charcoal has been made to burn with a brilliant white flame; water has been decomposed into its elementary parts; metals have been melted and set on fire; fragments of diamond, charcoal, and plumbago have been dispersed as if evaporated; platina, the hardest and the heaviest of the metals, has been melted as readily as wax in the flame of a candle; the sapphire, quartz, magnesia, lime, and the firmest compounds in nature, have been fused. Its effects on the animal system are no less surprising.

The agency of galvanism explains why porter has a different and more pleasant taste when drunk out of a pewter-pot than out of glass or earthenware; why works of metal which are soldered together soon tarnish in the place where the metals are joined; and why the copper sheathing of ships, when fastened with iron nails, is soon corroded about the place of contact. In all these cases a galvanic circle is formed which produces the effects.

THE SIAMESE TWINS GALVANISED.

It will be recollected that the Siamese twins, brought to England in the year 1829, were united by a jointed cartilaginous band. A silver tea spoon being placed on the tongue of one of the twins and a disc of zinc on the tongue of the other, the moment the two metals were brought into contact both the boys exclaimed, "Sour, sour;" thus proving that the galvanic influence passed from the one to the other through the connecting band.

MINUTE AND VAST BATTERIES.

Dr. Wollaston made a simple apparatus out of a silver thimble, with its top cut off. It was then partially flattened, and a small plate of zinc being introduced into it, the apparatus was immersed in a weak solution of sulphuric acid. With this minute battery, Dr. Wollaston was able to fuse a wire of platinum 1/3000th of an inch in diameter—a degree of tenuity to which no one had ever succeeded in drawing it.

Upon the same principle (that of introducing a plate of zinc between two plates of other metals) Mr. Children constructed his immense battery, the zinc plates of which measured six feet by two feet eight inches; each plate of zinc being placed between two of copper, and each triad of plates being enclosed in a separate cell. With this powerful apparatus a wire of platinum, 1/10th of an inch in diameter and upwards of five feet long, was raised to a red heat, visible even in the broad glare of daylight.

The great battery at the Royal Institution, with which Sir Humphry Davy discovered the composition of the fixed alkalies, was of immense power. It consisted of 200 separate parts, each composed of ten double plates, and each plate containing thirty-two square inches; the number of double plates being 2000, and the whole surface 128,000 square inches.

Mr. Highton, C.E., has made a battery which exposes a surface of only 1/100th part of an inch: it consists of but one cell; it is less than 1/10000th part of a cubic inch, and yet it produces electricity more than enough to overcome all the resistance in the inventor's brother's patent Gold-leaf Telegraph, and works the same powerfully. It is, in short, a battery which, although *it will go through the eye of a needle*, will yet work a telegraph well. Mr. Highton had previously constructed a battery in size less than

1/40th of a cubic inch: this battery, he found, would for a month together ring a telegraph-bell ten miles off.

ELECTRIC INCANDESCENCE OF CHARCOAL POINTS.

The most splendid phenomenon of this kind is the combustion of charcoal points. Pointed pieces of the residuum obtained from gas retorts will answer best, or Bunsen's composition may be used for this purpose. Put two such charcoal points in immediate contact with the wires of your battery; bring the points together, and they will begin to burn with a dazzling white light. The charcoal points of the large apparatus belonging to the Royal Institution became incandescent at a distance of 1/30th of an inch; when the distance was gradually increased till they were four inches asunder, they continued to burn with great intensity, and a permanent stream of light played between them. Professor Bunsen obtained a similar flame from a battery of four pairs of plates, its carbon surface containing 29 feet. The heat of this flame is so intense, that stout platinum wire, sapphire, quartz, talc, and lime are reduced by it to the liquid form. It is worthy of remark, that no combustion, properly so called, takes place in the charcoal itself, which sustains only an extremely minute loss in its weight and becomes rather denser at the points. The phenomenon is attended with a still more vivid brightness if the charcoal points are placed in a vacuum, or in any of those gases which are not supporters of combustion. Instead of two charcoal points, one only need be used if the following arrangement is adopted: lay the piece of charcoal on some quicksilver that is connected with one pole of the battery, and complete the circuit from the other pole by means of a strip of platinum. When Professor Peschel used a piece of well-burnt coke in the manner just described, he obtained a light which was almost intolerable to the eyes.

VOLTAIC ELECTRICITY.

On January 31, 1793, Volta announced to the Royal Society his discovery of the development of electricity in metallic bodies. Galvani had given the name of Animal Electricity to the power which caused spontaneous convulsions in the limbs of frogs when the divided nerves were connected by a metallic wire. Volta, however, saw the true cause of the phenomena described by Galvani. Observing that the effects were far greater when the connecting medium consisted of two different kinds of metal, he inferred that the principle of excitation existed in the metals, and not in the nerves of the animal; and he assumed that the exciting fluid was ordinary electricity, produced by the contact of the two metals; the convulsions of the frog consequently arose from the electricity thus developed passing along its nerves and muscles.

In 1800 Volta invented what is now called the Voltaic Pile, or compound Galvanic circle.

The term Animal Electricity (says Dr. Whewell) has been superseded by others, of which *Galvanism* is the most familiar; but I think that Volta's office in this discovery is of a much higher and more philosophical kind than that of Galvani; and it would on this account be more fitting to employ the term *Voltaic Electricity*, which, indeed, is very commonly used, especially by our most recent and comprehensive writers. The *Voltaic pile* was a more important step in the history of electricity than the Leyden jar had been—*Hist. Ind. Sciences*, vol. iii.

No one who wishes to judge impartially of the scientific history of these times and of its leaders, will consider Galvani and Volta as equals, or deny the vast superiority of the latter over all his opponents or fellow-workers, more especially over those of the Bologna school. We shall scarcely again find in one man gifts so rich and so calculated for research as were combined in Volta. He possessed that "incomprehensible talent," as Dove has called it, for separating the essential from the immaterial in complicated phenomena; that boldness of invention which must precede experiment, controlled by the most strict and cautious mode of manipulation; that unremitting attention which allows no circumstance to pass unnoticed; lastly, with so much acuteness, so much simplicity, so much grandeur of conception, combined with such depth of thought, he had a hand which was the hand of a workman.—*Jameson's Journal*, No. 106.

THE VOLTAIC BATTERY AND THE GYMNOTUS.

"We boast of our Voltaic Batteries," says Mr. Smee. "I should hardly be believed if I were to say that I did not feel pride in having constructed my own, especially when I consider the extensive operations which it has conducted. But when I compare my battery with the battery which nature has given to the electrical eel and the torpedo, how insignificant are human operations compared with those of the Architect of living beings! The stupendous electric eel in the Polytechnic Institution, when he seeks to kill his prey, encloses him in a circle; then, by volition, causes the voltaic force to be produced, and the hapless creature is instantly killed. It would probably require ten thousand of my artificial batteries to effect the same object, as the creature is killed *instanter* on receiving the shock. As much, however, as my battery is inferior to that of the electric fish, so is man superior to the same animal. Man is endowed with a power of mind competent to appreciate the force of matter, and is thus enabled to make the battery. The eel can but use the specific apparatus which nature has bestowed upon it."

Some observations upon the electric current around the gymnotus, and notes of experiments with this and other electric fish, will be found in *Things not generally Known*, p. 199.

VOLTAIC CURRENTS IN MINES.

Many years ago, Mr. R. W. Fox, from theory entertaining a belief that a connection existed between voltaic action in the interior of the earth and the arrangement of metalliferous veins, and also the progressive increase of temperature in the strata as we descend from the surface, endeavoured to verify the same from experiment in the mine of Huel Jewel, in Cornwall. His apparatus consisted of small plates of sheet-copper, which were fixed in contact with a plate in the veins with copper nails, or else wedged closely against them with wooden props stretched across the galleries. Between two of these plates, at different stations, a communication was made by means of a copper wire 1/20th of an inch in diameter, which included a galvanometer in its circuit. In some instances 300 fathoms of copper wire were employed. It was then found that the intensity of the voltaic current was generally greater in proportion to the greater abundance of copper ore in the veins, and in some degree to the depth of the stations. Hence Mr. Fox's discovery promised to be of practical utility to the miner in discovering the relative quantity of ore in the veins, and the directions in which it most abounds.

The result of extended experiments, mostly made by Mr. Robert Hunt, has not, however, confirmed Mr. Fox's views. It has been found that the voltaic currents detected in the lodes are due to the chemical decomposition going on there; and the more completely this process of decomposition is established, the more powerful are the voltaic currents. Meanwhile these have nothing whatever to do with the increase of temperature with depth. Recent observations, made in the deep mines of Cornwall under the direction of Mr. Fox, do not appear consistent with the law of thermic increase as formerly established, the shallow mines giving a higher ratio of increase than the deeper ones.

GERMS OF ELECTRIC KNOWLEDGE.

Two centuries and a half ago, Gilbert recognised that the property of attracting light substances when rubbed, be their nature what it may, is not peculiar to amber, which is a condensed earthy juice cast up by the waves of the sea, and in which flying insects, ants, and worms lie entombed as in eternal sepulchres. The force of attraction (Gilbert continues) belongs to a whole class of very different substances, as glass, sulphur, sealing-wax, and all resinous substances—rock crystal and all precious stones, alum and rock-salt. Gilbert measured the strength of the excited electricity by means of a small needle—not made of iron—which moved freely on a pivot, and perfectly similar to the apparatus used by Haüy and Brewster in testing the electricity excited in minerals by heat and friction. "Friction," says Gilbert further, "is productive of a stronger effect in dry than in humid air; and rubbing with silk cloths is most advantageous."

Otto von Guerike, the inventor of the air-pump, was the first who observed any thing more than mere phenomena of attraction. In his experiments with a rubbed piece of sulphur he recognised the phenomena of repulsion, which subsequently led to the establishment of the laws of the sphere of action and of the distribution of electricity. *He heard the first sound, and saw the first light, in artificially-produced electricity.* In an experiment instituted by Newton in 1675, the first traces of an electric charge in a rubbed plate of glass were seen.

TEMPERATURE AND ELECTRICITY.

Professor Tyndall has shown that all variations of temperature, in metals at least, excite electricity. When the wires of a galvanometer are brought in contact with the two ends of a heated poker, the prompt deflection of the galvanometer-needle indicates that a current of electricity has been sent through the instrument. Even the two ends of a spoon, one of which has been dipped in hot water, serve to develop an electric current; and in cutting a hot beefsteak with a steel knife and a silver fork there is an excitement of electricity. The mere heat of the finger is sufficient to cause the deflection of the galvanometer; and when ice is applied to the part that has been previously warmed, the galvanometer-needle is deflected in the contrary direction. A small instrument invented by Melloni is so extremely sensitive of the action of heat, that electricity is excited when the hand is held six inches from it.

VAST ARRANGEMENT OF ELECTRICITY.

Professor Faraday has shown that the Electricity which decomposes, and that which is evolved in the decomposition of, a certain quantity of matter, are alike. What an enormous quantity of electricity, therefore, is required for the decomposition of a single grain of water! It must be in quantity sufficient to sustain a platinum wire 1/104th of an inch in thickness red-hot in contact with the air for three minutes and three-quarters. It would appear that 800,000 charges of a Leyden battery, charged by thirty turns of a very large and powerful plate-machine in full action, are necessary to supply electricity sufficient to decompose a single grain of water, or to equal the quantity of electricity which is naturally associated with the elements of that grain of water, endowing them with their mutual chemical affinity. Now the above quantity of electricity, if passed at once through the head of a rat or a cat, would kill it as by a flash of lightning. The quantity is, indeed, equal to that which is developed from a charged thunder-cloud.

DECOMPOSITION OF WATER BY ELECTRICITY.

Professor Andrews, by an ingenious arrangement, is enabled to show that water is decomposed by the common machine; and by using an electrical kite, he was able, in fine weather, to produce decomposition, although so slowly that only 1/700000th of a grain of water was decomposed per hour. Faraday has proved that the decomposition of one single grain of water produces more electricity than is contained in the most powerful flash of lightning.

ELECTRICITY IN BREWING.

Mr. Black, a practical writer upon Brewing, has found that by the practice of imbedding the fermentation-vats in the earth, and connecting them by means of metallic pipes, an electrical current passes through the beer and causes it to turn sour. As a preventive, he proposed to place the vats upon wooden blocks, or on any other non-conductors, so that they may be insulated. It has likewise been ascertained that several brewers who had brewed excellent ale on the south side of the street, on removing to the north have failed to produce good ale.

ELECTRIC PAPER.

Professor Schonbein has prepared paper, as transparent as glass and impermeable to water, which develops a very energetic electric force. By placing some sheets on each other, and simply rubbing them once or twice with the hand, it becomes difficult to separate them. If this experiment is performed in the dark, a great number of distinct flashes may be perceived between the separated surfaces. The disc of the electrophorus, placed on a sheet that has been rubbed, produces sparks of some inches in length. A thin and very dry sheet of paper, placed against the wall, will adhere strongly to it for several hours if the hand be passed only once over it. If the same sheet be passed between the thumb and fore-finger in the dark, a luminous band will be visible. Hence with this paper may be made powerful and cheap electrical machines.

DURATION OF THE ELECTRIC SPARK.

By means of Professor Wheatstone's apparatus, the Duration of the Electric Spark has been ascertained not to exceed the twenty-five-thousandth part of a second. A cannon-ball, if illumined in its flight by a flash of lightning, would, in consequence of the momentary duration of the light, appear to be stationary, and even the wings of an insect, that move ten thousand times in a second, would seem at rest.

VELOCITY OF ELECTRIC LIGHT.

On comparing the velocities of solar, stellar, and terrestrial light, which are all equally refracted in the prism, with the velocity of the light of frictional electricity, we are disposed, in accordance with Wheatstone's ingeniously-conducted experiments, to regard the lowest ratio in which the latter excels the former as 3:2. According to the lowest results of Wheatstone's apparatus, electric light traverses 288,000 miles in a second. If we reckon 189,938 miles for stellar light, according to Struve, we obtain the difference of 95,776 miles as the greater velocity of electricity in one second.

From the experiment described in Wheatstone's paper (*Philosophical Transactions* for 1834), it would appear that the human eye is capable of perceiving phenomena of light whose duration is limited to the millionth part of a second.

In Professor Airy's experiments with the electric telegraph to determine the difference of longitude between Greenwich and Brussels, the time spent by the electric current in passing from one observatory to the other (270 miles) was found to be 0•109″ or rather more than *the ninth part of a second*; and this determination rests on 2616 observations: a speed which would "girdle the globe" in ten seconds.

IDENTITY OF ELECTRIC AND MAGNETIC ATTRACTION.

This vague presentiment of the ancients has been verified in our own times. "When electrum (amber)," says Pliny, "is animated by friction and heat, it will attract bark and dry leaves precisely as the loadstone attracts iron." The same words may be found in the literature of an Asiatic nation, and occur in a eulogium on the loadstone by the Chinese physicist Knopho, in the fourth century: "The magnet attracts iron as amber does the smallest grain of mustard-seed. It is like a breath of wind, which mysteriously penetrates through both, and communicates itself with the rapidity of an arrow."

Humboldt observed with astonishment on the woody banks of the Orinoco, in the sports of the natives, that the excitement of electricity by friction was known to these savage races. Children may be seen to rub the dry, flat, and shining seeds or husks of a trailing plant until they are able to attract threads of cotton and pieces of bamboo-cane. What a chasm divides the electric pastime of these naked copper-coloured Indians from the discovery of a metallic conductor discharging its electric shocks, or a pile formed of many chemically-decomposing substances, or a light-engendering magnetic apparatus! In such a chasm lie buried thousands of years, that compose the history of the intellectual development of mankind.— *Humboldt's Cosmos*, vol. i.

THEORY OF THE ELECTRO-MAGNETIC ENGINE.

Several years ago a speculative American set the industrial world of Europe in excitement by this proposition. The Magneto-Electric Machines often made use of in the case of rheumatic disorders are well known. By imparting a swift rotation to the magnet of such a machine, we obtain powerful currents of electricity. If these be conducted through water, the latter will be reduced to its two components, oxygen and hydrogen. By the combustion of hydrogen water is again generated. If this combustion takes place, not in atmospheric air, in which oxygen only constitutes a fifth part, but in pure oxygen, and if a bit of chalk be placed in the flame, the chalk will be raised to a white heat, and give us the sun-like Drummond light: at the same time the flame develops a considerable quantity of heat. Now the American inventor proposed to utilise in this way the gases obtained from electrolytic decomposition; and asserted that by the combustion a sufficient amount of heat was generated to keep a small steam-engine in action, which again drove his magneto-electric machine, decomposed the water, and thus continually prepared its own fuel. This would certainly have been the most splendid of all discoveries,—a perpetual motion which, besides the force that kept it going, generated light like the sun, and warmed all around it. The affair, however, failed, as was predicted by those acquainted with the physical investigations which bear upon the subject.—*Professor Helmholtz.*

MAGNETIC CLOCK AND WATCH.

In the Museum of the Royal Society are two curiosities of the seventeenth century which are objects of much interest in association with the electric discoveries of our day. These are a Clock, described by the Count Malagatti (who accompanied Cosmo III., Grand Duke of Tuscany, to inspect the Museum in 1669) as more worthy of observation than all the other objects in the cabinet. Its "movements are derived from the vicinity of a loadstone, and it is so adjusted as to discover the distance of countries at sea by the longitude." The analogy between this clock and the electric clock of the present day is very remarkable. Of kindred interest is "Hook's Magnetic Watch," often alluded to in the Royal Society's Journal-book of 1669 as "going slower or faster according to the greater or less distance of the loadstone, and so moving regularly in any posture."

WHEATSTONE'S ELECTRO-MAGNETIC CLOCK.

In this ingenious invention, the object of Professor Wheatstone was to enable a simple clock to indicate exactly the same time in as many different places, distant from each other, as may be required. A standard clock in an observatory, for example, would thus keep in order another clock in each apartment, and that too with such accuracy, that *all of them, however numerous, will beat dead seconds audibly with as great precision as the standard astronomical time-piece with which they are connected.* But, besides this, the subordinate time-pieces thus regulated require none of the mechanism for maintaining or regulating the power. They consist simply of a face, with its second, minute, and hour hands, and a train of wheels which communicate motion from the action of the second-hand to that of the hour-hand, in the same manner as an ordinary clock-train. Nor is this invention confined to observatories and large establishments. The great horologe of St. Paul's might, by a suitable network of wires, or even by the existing metallic pipes of the metropolis, be made to command and regulate all the other steeple-clocks in the city, and even every clock within the precincts of its metallic bounds. As railways and telegraphs extend from London nearly to the remotest cities and villages, the sensation of time may be transmitted along with the elements of language; and the great cerebellum of the metropolis may thus constrain by its sympathies, and regulate by its power, the whole nervous system of the empire.

HOW TO MAKE A COMMON CLOCK ELECTRIC.

M. Kammerer of Belgium effects this by an addition to any clock whereby it is brought into contact with the two poles of a galvanic battery, the wires from which communicate with a drum moved by the clockwork; and every fifteen seconds the current is changed, the positive and the negative being transmitted alternately. A wire is continued from the drum to the electric clock, the movement of which, through the plate-glass dial, is seen to be two pairs of small straight electro-magnets, each pair having their ends opposite to the other pair, with about half an inch space between. Within this space there hangs a vertical steel bar, suspended from a spindle at the top. The rod has two slight projections on each side parallel to the ends of the wire-coiled magnets. When the electric current comes on the wire from the positive end of the battery (through the drum of the regulator-clock) the positive magnets attract the bar to it, the distance being perhaps the sixteenth of an inch. When, at the end of fifteen seconds, the negative pole operates, repulsion takes effect, and the bar moves to the opposite side. This oscillating bar gives motion to a wheel which turns the minute and hour hands.

M. Kammerer states, that if the galvanic battery be attached to any particular standard clock, any number of clocks, wherever placed, in a city or kingdom, and communicating with this by a wire, will indicate precisely the same time. Such is the precision, that the sounds of three clocks thus beating simultaneously have been mistaken as proceeding from one clock.

DR. FRANKLIN'S ELECTRICAL KITE.

Several philosophers had observed that lightning and electricity possessed many common properties; and the light which accompanied the explosion, the crackling noise made by the flame, and other phenomena, made them suspect that lightning might be electricity in a highly powerful state. But this connection was merely the subject of conjecture until, in the year 1750, Dr. Franklin suggested an experiment to determine the question. While he was waiting for the building of a spire at Philadelphia, to which he intended to attach his wire, the experiment was successfully made at Marly-la-Ville, in France, in the year 1752; when lightning was actually drawn from the clouds by means of a pointed wire, and it was proved to be really the electric fluid.

Almost every early electrical discovery of importance was made by Fellows of the Royal Society, and is to be found recorded in the *Philosophical Transactions*. In the forty-fifth volume occurs the first mention of Dr. Franklin's name, and his theory of positive and negative electricity. In 1756 he was elected into the Society, "without any fee or other payment." His previous communications to the *Transactions*, particularly the account of his electrical kite, had excited great interest. (*Weld's History of the Royal Society.*) It is thus described by him in a letter dated Philadelphia, October 1, 1752:

"As frequent mention is made in the public papers from Europe of the success of the Marly-la-Ville experiment for drawing the electric fire from clouds by means of pointed rods of iron erected on high buildings, &c., it may be agreeable to the curious to be informed that the same experiment has succeeded in Philadelphia, though made in a different and more easy manner, which any one may try, as follows:

Make a small cross of two light strips of cedar, the arms so long as to reach to the four corners of a large thin silk handkerchief when extended. Tie the corners of the handkerchief to the extremities of the cross; so you have the body of a kite, which, being properly accommodated with a tail, loop, and string, will rise in the air like a kite made of paper; but this, being of silk, is fitter to bear the wet and wind of a thunder-gust without tearing. To the top of the upright stick of the cross is to be fixed a very sharp-pointed wire, rising a foot or more above the wood. To the end of the twine, next the band, is to be tied a silk ribbon; and where the twine and silk join a key may be fastened.

The kite is to be raised when a thunder-gust appears to be coming on, and the person who holds the string must stand within a door or window, or under some cover, so that the silk ribbon may not be wet; and care must be taken that the twine does not touch the frame of the door or window. As soon as any of the thunder-clouds come over the kite, the pointed wire will draw the electric fire from them; and the kite, with all the twine, will be electrified; and the loose filaments of the twine will stand out every way, and be attracted by an approaching finger.

When the rain has wet the kite and twine, so that it can conduct the electric fire freely, you will find it stream out plentifully from the key on the approach of your knuckle. At this key the phial may be charged; and from electric fire thus obtained spirits may be kindled, and all the other electrical experiments be performed which are usually done by the help of a rubbed-glass globe or tube; and thus the sameness of the electric matter with that of lightning is completely demonstrated."—*Philosophical Transactions.*

Of all this great man's (Franklin's) scientific excellencies, the most remarkable is the smallness, the simplicity, the apparent inadequacy of the means which he employed in his experimental researches. His discoveries were all made with hardly any apparatus at all; and if at any time he had been led to employ instruments of a somewhat less ordinary description, he never rested satisfied until he had, as it were, afterwards translated the process by resolving the problem with such simple machinery that you might say he had done it wholly unaided by apparatus. The experiments by which the identity of lightning and electricity was demonstrated were made with a sheet of brown paper, a bit of twine or silk thread, and an iron key!—*Lord Brougham.*[50]

FATAL EXPERIMENT WITH LIGHTNING.

These experiments are not without danger; and a flash of lightning has been found to be a very unmanageable instrument. In 1753, M. Richman, at St. Petersburg, was making an experiment of this kind by drawing lightning into his room, when, incautiously bringing his head too near the wire, he was struck dead by the flash, which issued from it like a globe of blue fire, accompanied by a dreadful explosion.

FARADAY'S ELECTRICAL ILLUSTRATIONS.

The following are selected from the very able series of lectures delivered by Professor Faraday at the Royal Institution:

The Two Electricities.—After having shown by various experiments the attractions and repulsions of light substances from excited glass and from an excited tube of gutta-percha, Professor Faraday proceeds to point out the difference in the character of the electricity produced by the friction of the two substances. The opposite characters of the electricity evolved by the friction of glass and of that excited by the friction of gutta-percha and shellac are exhibited by several experiments, in which the attraction of the positive and negative electricities to each other and the neutralisation of electrical action on the combination of the two forces are distinctly observable. Though adopting the terms "positive" and "negative" in distinguishing the electricity excited by glass from that excited by gutta-percha and resinous bodies, Professor Faraday is strongly opposed to the Franklinian theory from which these terms are derived. According to Franklin's view of the nature of electrical excitement, it arises from the disturbance, by friction or other means, of the natural quantity of one electric fluid which is possessed by all bodies; an excited piece of glass having more than its natural share, which has been taken from the rubber, the latter being consequently in a minus or negative state. This theory Professor Faraday considers to be opposed to the distinct characteristic actions of the two forces; and, in his opinion, it is impossible to deprive any body of electricity, and reduce it to the minus state of Franklin's hypothesis. Taking a Zamboni's pile, he applies its two ends separately to an electrometer, to show that each end produces opposite kinds of electricity, and that the zero, or absence of electrical excitement, only exists in the centre of the pile. To prove how completely the two electricities neutralise each other, an excited rod of gutta-percha and the piece of flannel with which it has been rubbed are laid on the top of the electrometer without any sign of electricity whilst they are together; but when either is removed, the gold leaves diverge with positive and negative electricity alternately. The Professor dwells strongly on the peculiarity of the dual force of electricity, which, in respect of its duality, is unlike any other force in nature. He then contrasts its phenomena of instantaneous conduction with those of the somewhat analogous force of heat; and he illustrates by several striking experiments the peculiar property which static electricity possesses of being spread only over the surfaces of bodies. A metal ice-pail is placed on an insulated stand and electrified, and a metal ball suspended by a string is introduced, and touches the bottom and sides without having any electricity imparted to it, but on touching the outside it becomes strongly electrical. The experiment is repeated with a wooden tub with the same result; and Professor Faraday mentions the still more remarkable manner in which he has proved the surface distribution of electricity by having a small chamber constructed and covered with tinfoil, which can be insulated; and whilst torrents of electricity are being evolved from the external surface, he enters it with a galvanometer, and cannot perceive the slightest manifestation of electricity within.

The Two Threads.—A curious experiment is made with two kinds of thread used as the conducting force. From the electric machine on the table a silk thread is first carried to the indicator a yard or two off, and is shown to be a non-conductor when the glass tube is rubbed and applied to the machine (although the silk, when wetted, conducted); while a metallic thread of the same thickness, when treated in the same way, conducts the force so much as to vehemently agitate the gold leaves within the indicator.

Non-conducting Bodies.—The action that occurs in bodies which cannot conduct is the most important part of electrical science. The principle is illustrated by the attraction and repulsion of an electrified ball of gilt paper by a glass tube, between which and the ball a sheet of shellac is suspended. The nearer a ball of another description—an unelectrical insulated body—is brought to the Leyden jar when charged, the greater influence it is seen to possess over the gold leaf within the indicator, by induction, not by conduction. The questions, how electricities attract each other, what kind of electricity is drawn from the machine to the hand, how the hand was electric, are thus illustrated. To show the divers operations of this wonderful force, a tub (a bad conductor) is placed by the electric machine. When the latter is charged, a ball, having been electrified from it, is held in the tub, and rattles against its sides and bottom. On the application of the ball to the indicator, the gold leaf is shown not to move, whereas it is agitated manifestly when the same process is gone through with the exception that the ball is made to touch the outside only of the tub. Similar experiments with a ball in an ice-pail and a vessel of wire-gauze, into the latter of which is introduced a mouse, which is shown to receive no shock, and not to be frightened at all; while from the outside of the vessel electric sparks are rapidly produced. This latter demonstration proves that, as the mouse, so men and women, might be safe inside a building with proper conductors while lightning played about the exterior. The wire-gauze being turned inside out, the principle is shown to be irreversible in spite of the change—what has been the unelectrical inside of the vessel being now, when made the outside portion, capable of receiving and transmitting the power, while the original outside is now unelectrical.

Repulsion of Bodies.—A remarkable and playful experiment, by which the repulsion of bodies similarly electrified is illustrated, consists in placing a basket containing a heap of small pieces of paper on an insulated stand, and connecting it with the prime conductor of the electrical machine; when the pieces of paper rise rapidly after each other into the air, and descend on the lecture-table like a fall of snow. The effect is greatly increased when a metal disc is substituted for the basket.

ORIGIN OF THE LEYDEN JAR.

Muschenbroek and Linnæus had made various experiments of a strong kind with water and wire. The former, as appears from a letter of his to Réaumur, filled a small bottle with water, and having corked it up, passed a wire through the cork into the bottle. Having rubbed the vessel on the outside and suspended it to the electric machine, he was surprised to find that on trying to pull the wire out he was subjected to an awfully severe shock in his joints and his whole body, such as he declared he would not suffer again for any experiment. Hence the Leyden jar, which owes its name to the University of Leyden, with which, we believe, Muschenbroek was connected.—*Faraday.*

DANGER TO GUNPOWDER MAGAZINES.

By the illustration of a gas globule, which is ignited from a spark by induction, Mr. Faraday has proved in a most interesting manner that the corrugated-iron roofs of some gunpowder-magazines,—on the subject of which he had often been consulted by the builders, with a view to the greater safety of these manufactories,—are absolutely dangerous by the laws of induction; as, by the return of induction, while a storm was discharging itself a mile or two off, a secondary spark might ignite the building.

ARTIFICIAL CRYSTALS AND MINERALS.—"THE CROSSE MITE."

Among the experimenters on Electricity in our time who have largely contributed to the "Curiosities of Science," Andrew Crosse is entitled to special notice. In his school-days he became greatly attached to the study of electricity; and on settling on his paternal estate, Fyne Court, on the Quantock Hills in Somersetshire, he there devoted himself to chemistry, mineralogy, and electricity, pursuing his experiments wholly independently of theories, and searching only for facts. In Holwell Cavern, near his residence, he observed the sides and the roof covered with Arragonite crystallisations, when his observations led him to conclude that the crystallisations were the effects, at least to some extent, of electricity. This induced him to make the attempt to form artificial crystals by the same means, which he began in 1807. He took some water from the cave, filled a tumbler, and exposed it to the action of a voltaic battery excited by water alone, letting the platinum-wires of the battery fall on opposite sides of the tumbler from the opposite poles of the battery. After ten days' constant action, he produced crystals of carbonate of lime; and on repeating the experiment in the dark, he produced them in six days. Thus Mr. Crosse simulated in his laboratory one of the hitherto most mysterious processes of nature.

He pursued this line of research for nearly thirty years at Fyne Court, where his electrical-room and laboratory were on an enormous scale: the apparatus had cost some thousands of pounds, and the house was nearly full of furnaces. He carried an insulated wire above the tops of the trees around his house to the length of a mile and a quarter, afterwards shortened to 1800 feet. By this wire, which was brought into connection with the apparatus in a chamber, he was enabled to see continually the changes in the state of the atmosphere, and could use the fluid so collected for a variety of purposes. In 1816, at a meeting of country gentlemen, he prophesied that, "by means of electrical agency, we shall be able to communicate our thoughts simultaneously with the uttermost ends of the earth." Still, though he foresaw the powers of the medium, he did not make any experiments in that direction, but confined himself to the endeavour to produce crystals of various kinds. He ultimately obtained forty-one mineral crystals, or minerals uncrystallised, in the form in which they are produced by nature, including one sub-sulphate of copper—an entirely new mineral, neither found in nature nor formed by art previously. His belief was that even diamonds might be produced in this way.

Mr. Crosse worked alone in his retreat until 1836, when, attending the meeting of the British Association at Bristol, he was induced to explain his experiments, for which he was highly complimented by Dr. Buckland, Dr. Dalton, Professor Sedgwick, and others.51

Shortly after Mr. Crosse's return to Fyne Court, while pursuing his experiments for forming crystals from a highly caustic solution out of contact with atmospheric air, he was greatly surprised by the appearance of an insect. Black flint, burnt to redness and reduced to powder, was mixed with carbonate of potash, and exposed to a strong heat for fifteen minutes; and the mixture was poured into a black-lead crucible in an air furnace. It was reduced to powder while warm, mixed with boiling water, kept boiling for some minutes, and then hydrochloric acid was added to supersaturation. After being exposed to voltaic action for twenty-six days, a perfect insect of the Acari tribe made its appearance, and in the course of a few weeks about a hundred more. The experiment was repeated in other chemical fluids with the like results; and Mr. Weeks of Sandwich afterwards produced the Acari inferrocyanerret of potassium. The Acarus of Mr. Crosse was found to contribute a new species of that genus, nearly approaching the Acari found in cheese and flour, or more nearly, Hermann's *Acarus dimidiatus*.

This discovery occasioned great excitement. The possibility was denied, though Mr. Faraday is said to have stated in the same year that he had seen similar appearances in his own electrical experiments. Mr. Crosse was now accused of impiety and aiming at creation, to which attacks he thus replied:

As to the appearance of the acari under long-continued electrical action, I have never in thought, word, or deed given any one a right to suppose that I considered them as a creation, or even as a formation, from inorganic matter. To create is to form a something out of a nothing. To annihilate is to reduce that something to a nothing. Both of these, of course, can only be the attributes of the Almighty. In fact, I can assure you most sacredly that I have never dreamed of any theory sufficient to account for their appearance. I confess that I was not a little surprised, and am so still, and quite as much as I was when the acari made their first appearance. Again, I have never claimed any merit as attached to these experiments. It was a matter of chance; I was looking for silicious formations, and animal matter appeared instead.

These Acari, if removed from their birthplace, lived and propagated; but uniformly died on the first recurrence of frost, and were entirely destroyed if they fell back into the fluid whence they arose.

One of Mr. Crosse's visitors thus describes the vast electrical room at Fyne Court:

Here was an immense number of jars and gallipots, containing fluids on which electricity was operating for the production of crystals. But you are startled in the midst of your observations by the smart crackling sound that attends the passage of the electrical spark; you hear also the rumbling of distant thunder. The rain is already plashing in great drops against the glass, and the sound of the passing sparks continues to startle your ear; you see at the window a huge brass conductor, with a discharging rod near it passing into the floor, and from the one knob to the other sparks are leaping with increasing rapidity and noise, every one of which would kill twenty men at one blow, if they were linked together hand in hand and the spark sent through the circle. From this conductor wires pass off without the window, and the electric fluid is conducted harmlessly away. Mr. Crosse approached the instrument as boldly as if the flowing stream of fire were a harmless spark. Armed with his insulated rod, he sent it into his batteries: having charged them, he showed how wire was melted, dissipated in a moment, by its passage; how metals—silver, gold, and tin—were inflamed and burnt like paper, only with most brilliant hues. He showed you a mimic aurora and a falling-star, and so proved to you the cause of those beautiful phenomena.

Mr. Crosse appears to have produced in all "about 200 varieties of minerals, exactly resembling in all respects similar ones found in nature." He tried also a new plan of extracting gold from its ores by an electrical process, which succeeded, but was too expensive for common use. He was in the habit of saying that he could, like

Archimedes, move the world "if he were able to construct a battery at once cheap, powerful, and durable." His process of extracting metals from their ores has been patented. Among his other useful applications of electricity are the purifying by its means of brackish or sea-water, and the improving bad wine and brandy. He agreed with Mr. Quekett in thinking that it is by electrical action that silica and other mineral substances are carried into and assimilated by plants. Negative electricity Mr. Crosse found favourable to no plants except fungi; and positive electricity he ascertained to be injurious to fungi, but favourable to every thing else.

Mr. Crosse died in 1855. His widow has published a very interesting volume of *Memorials* of the ingenious experimenter, from which we select the following:

On one occasion Mr. Crosse kept a pair of soles under the electric action for three months; and at the end of that time they were sent to a friend, whose domestics knew nothing of the experiment. Before the cook dressed them, her master asked her whether she thought they were fresh, as he had some doubts. She replied that she was sure they were fresh; indeed, she said she could swear that they were alive yesterday! When served at table they appeared like ordinary fish; but when the family attempted to eat them, they were found to be perfectly tasteless—the electric action had taken away all the essential oil, leaving the fish unfit for food. However, the process is exceedingly useful for keeping fish, meat, &c. fresh and *good* for ten days or a fortnight. I have never heard a satisfactory explanation of the cause of the antiseptic power communicated to water by the passage of the electric current. Whether ozone has not something to do with it, may be a question. The same effect is produced whichever two dissimilar metals are used.

The Electric Telegraph.

ANTICIPATIONS OF THE ELECTRIC TELEGRAPH.

The great secret of ubiquity, or at least of instantaneous transmission, has ever exercised the ingenuity of mankind in various romantic myths; and the discovery of certain properties of the loadstone gave a new direction to these fancies.

The earliest anticipation of the Electric Telegraph of this purely fabulous character forms the subject of one of the *Prolusiones Academicæ* of the learned Italian Jesuit Strada, first published at Rome in the year 1617. Of this poem a free translation appeared in 1750. Strada's fancy was this: "There is," he supposes, "a species of loadstone which possesses such virtue, that if two needles be touched with it, and then balanced on separate pivots, and the one be turned in a particular direction, the other will sympathetically move parallel to it. He then directs each of these needles to be poised and mounted parallel on a dial having the letters of the alphabet arranged round it. Accordingly, if one person has one of the dials, and another the other, by a little pre-arrangement as to details a correspondence can be maintained between them at any distance by simply pointing the needles to the letters of the required words.

Strada, in his poetical reverie, dreamt that some such sympathy might one day be found to hold up the Magnesian Stone."

Strada's conceit seems to have made a profound impression on the master-minds of the day. His poem is quoted in many works of the seventeenth and eighteenth centuries; and Bishop Wilkins, in his book on Cryptology, is strangely afraid lest his readers should mistake Strada's fancy for fact. Wilkins writes: "This invention is altogether imaginary, having no foundation in any real experiment. You may see it frequently confuted in those that treat concerning magnetical virtues."

Again, Addison, in the 241st No. of the *Spectator*, 1712, describes Strada's "Chimerical correspondence," and adds that, "if ever this invention should be revived or put in practice," he "would propose that upon the lover's dial-plate there should be written not only the four-and-twenty letters, but several entire words which have always a place in passionate epistles, as flames, darts, die, language, absence, Cupid, heart, eyes, being, drown, and the like. This would very much abridge the lover's pains in this way of writing a letter, as it would enable him to express the most useful and significant words with a single touch of the needle."

After Strada and his commentators comes Henry Van Etten, who shows how "Claude, being at Paris, and John at Rome, might converse together, if each had a needle touched by a stone of such virtue that as one moved itself at Paris the other should be moved at Rome:" he adds, "it is a fine invention, but I do not think there is a magnet in the world which has such virtue; besides, it is inexpedient, for treasons would be too frequent and too much protected. (*Recréations Mathématiques*: see 5th edition, Paris, 1660, p. 158.) Sir Thomas Browne refers to this "conceit" as "excellent, and, if the effect would follow, somewhat divine;" but he tried the two needles touched with the same loadstone, and placed in two circles of letters, "one friend keeping one and another the other, and agreeing upon an hour when they will communicate," and found the tradition a failure that, "at what distance of place soever, when one needle shall be removed unto any letter, the other, by a wonderful sympathy, will move unto the same." (See *Vulgar Errors*, book ii. ch. iii.)

Glanvill's *Vanity of Dogmatizing*, a work published in 1661, however, contains the most remarkable allusion to the prevailing telegraphic fancy. Glanvill was an enthusiast, and he clearly predicts the discovery and general adoption of the electric telegraph. "To confer," he says, "at the distance of the Indies by sympathetic conveyance may be as usual to future times as to us in a literary correspondence." By the word "sympathetic" he evidently intended to convey magnetic agency; for he subsequently treats of "conference at a distance by impregnated needles," and describes the device substantially as it is given by Sir Thomas Browne, adding, that though it did not then answer, "by some other such way of magnetic efficiency it may hereafter with success be attempted, when magical history shall be enlarged by riper inspection; and 'tis not unlikely but that present discoveries might be improved to the performance." This may be said to close the most speculative or mythical period in reference to the subject of electro-telegraphy.

Electricians now began to be sedulous in their experiments upon the new force by friction, then the only known method of generating electricity. In 1729, Stephen Gray, a pensioner of the Charter-house, contrived a method of making electrical signals through a wire 765 feet long; yet this most important experiment did not excite much attention. Next Dr. Watson, of the Royal Society, experimented on the possibility of transmitting electricity through a large circuit from the simple fact of Le Monnier's account of his feeling the stroke of the electrified fires through two of the basins of the Tuileries (which occupy nearly an acre), by means of an iron chain lying upon the ground and stretched round half their circumference. In 1745, Dr. Watson, assisted by several members of the Royal Society, made a series of experiments to ascertain how far electricity could be conveyed by means of conductors. "They caused the shock to pass across the Thames at Westminster Bridge, the circuit being completed by making use of the river for one part of the chain of communication. One end of the wire communicated with the coating of a charged phial, the other being held by the observer, who in his other hand held an iron rod which he dipped into the river. On the opposite side of the river stood a gentleman, who likewise dipped an iron rod in the river with one hand, and in the other held a wire the extremity of which might be brought into contact with the wire of the phial. Upon making the discharge, the shock was felt simultaneously by both the observers." (*Priestley's History of Electricity.*) Subsequently the same parties made experiments near Shooter's Hill, when the wires formed a circuit of four miles, and conveyed the shock with equal facility,—"a distance which without trial," they observed, "was too great to be credited."[52] These experiments in 1747 established two great principles: 1, that the electric current is transmissible along nearly two miles and a half of iron wire; 2, that the electric current may be completed by burying the poles in the earth at the above distance.

In the following year, 1748, Benjamin Franklin performed his celebrated experiments on the banks of the Schuylkill, near Philadelphia; which being interrupted by the hot weather, they were concluded by a picnic, when spirits were fired by an electric spark sent through a wire in the river, and a turkey was killed by the electric shock, and roasted by the electric jack before a fire kindled by the electrified bottle.

In the year 1753, there appeared in the *Scots' Magazine*, vol. xv., definite proposals for the construction of an electric telegraph, requiring as many conducting wires as there are letters in the alphabet; it was also proposed to converse by chimes, by substituting bells for the balls. A similar system of telegraphing was next invented by Joseph Bozolus, a Jesuit, at Rome; and next by the great Italian electrician Tiberius Cavallo, in his treatise on Electricity.

In 1787, Arthur Young, when travelling in France, saw a model working telegraph by M. Lomond: "You write two or three words on a paper," says Young; "he takes it with him into a room, and turns a machine enclosed in a cylindrical case, at the top of which is an electrometer—a small fine pith-ball; a wire connects with a similar cylinder and electrometer in a distant apartment; and his wife, by remarking the corresponding motions of the ball, writes down the words they indicate: from which it appears that he has formed an alphabet of motions. As the length of the wire makes no difference in the effect, a correspondence might be carried on at any distance. Whatever the use may be, the invention is beautiful."

We now reach a new epoch in the scientific period—the discovery of the Voltaic Pile. In 1794, according to *Voigt's Magazine*, Reizen made use of the electric spark for the telegraph; and in 1798 Dr. Salva of Madrid constructed a similar telegraph, which the Prince of Peace subsequently exhibited to the King of Spain with great success.

In 1809, Soemmering exhibited a telegraphic apparatus worked by galvanism before the Academy of Sciences at Munich, in which the mode of signalling consisted in the development of gas-bubbles from the decomposition of water placed in a series of glass tubes, each of which denoted a letter of the alphabet. In 1813, Mr. Sharpe, of Doe Hill near Alfreton, devised a *voltaic*-electric telegraph, which he exhibited to the Lords of the Admiralty, who spoke approvingly of it, but declined to carry it into effect. In the following year, Soemmering exhibited a *voltaic*-electric telegraph of his own construction, which, however, was open to the objection of there being as many wires as signs or letters of the alphabet.

The next invention is of much greater importance. Upon the suggestion of Cavallo, already referred to, Francis Ronalds constructed a perfect electric telegraph, employing frictional electricity notwithstanding Volta's discoveries had been known in England for sixteen years. This telegraph was exhibited at Hammersmith in

1816:53 it consisted of a single insulated wire, the indication being by pith-balls in front of a dial. When the wire was charged, the balls were divergent, but collapsed when the wire was discharged; at the same time were employed two clocks, with lettered discs for the signals. "If, as Paley asserts (and Coleridge denies), 'he alone discovers who proves,' Ronalds is entitled to the appellation of the first discoverer of an efficient electric telegraph." (*Saturday Review*, No. 14754) Nevertheless the Government of the day refused to avail itself of this admirable contrivance.

In 1819, Oersted made his great discovery of the deflection, by a current of electricity, of a magnetic needle at right angles to such current. Dr. Hamel of St. Petersburg states that Baron Schilling was the first to apply Oersted's discovery to telegraphy; Ampère had previously suggested it, but his plan was very complicated, and Dr. Hamel maintains that Schilling first realised the idea by actually producing an electro-magnetic telegraph simpler in construction than that which Ampère had *imagined*. In 1836, Professor Muncke of Heidelberg, who had inspected Schilling's telegraphic apparatus, explained the same to William Fothergill Cooke, who in the following year returned to England, and subsequently, with Professor Wheatstone, laboured simultaneously for the introduction of the electro-magnetic telegraph upon the English railways; the first patent for which was taken out in the joint names of these two gentlemen.

In 1844, Professor Wheatstone, with one of his telegraphs, formed a communication between King's College and the lofty shot-tower on the opposite bank of the Thames: the wire was laid along the parapets of the terrace of Somerset House and Waterloo Bridge, and thence to the top of the tower, about 150 feet high, where a telegraph was placed; the wire then descended, and a plate of zinc attached to its extremity was plunged into the mud of the river, whilst a similar plate attached to the extremity at the north side was immersed in the water. The circuit was thus completed by the entire breadth of the Thames, and the telegraph acted as well as if the circuit were entirely metallic.

Shortly after this experiment, Professor Wheatstone and Mr. Cooke laid down the first working electric telegraph on the Great Western Railway, from Paddington to Slough.

ELECTRIC GIRDLE FOR THE EARTH.

One of our most profound electricians is reported to have exclaimed: "Give me but an unlimited length of wire, with a small battery, and I will girdle the universe with a sentence in forty minutes." Yet this is no vain boast; for so rapid is the transition of the electric current along the line of the telegraph wire, that, supposing it were possible to carry the wires eight times round the earth, the transit would occupy but *one second of time*!

CONSUMPTION OF THE ELECTRIC TELEGRAPH.

It is singular to see how this telegraphic agency is measured by the chemical consumption of zinc and acid. Mr. Jones (who has written a work upon the Electric Telegraphs of America) estimates that to work 12,000 miles of telegraph about 3000 zinc cups are used to hold the acid: these weigh about 9000 lbs., and they undergo decomposition by the galvanic action in about six months, so that 18,000 lbs. of zinc are consumed in a year. There are also about 3600 porcelain cups to contain nitric acid; it requires 450 lbs. of acid to charge them once, and the charge is renewed every fortnight, making about 12,000 lbs. of nitric acid in a year.

TIME LOST IN ELECTRIC MESSAGES.

Although it may require an hour, or two or three hours, to transmit a telegraphic message to a distant city, yet it is the mechanical adjustment by the sender and receiver which really absorbs this time; the actual transit is practically instantaneous, and so it would be from here to the antipodes, so far as the current itself is concerned.

THE ELECTRIC TELEGRAPH IN ASTRONOMY AND THE DETERMINATION OF LONGITUDE.

The Electric Telegraph has become an instrument in the hands of the astronomer for determining the difference of longitude between two observatories. Thus in 1854 the difference of longitude between London and Paris was determined within a limit of error which amounted barely to a quarter of a second. The sudden disturbances of the magnetic needle, when freely suspended, which seem to take place simultaneously over whole continents, if not over the whole globe, from some unexplained cause, are pointed out as means by which the differences of longitude between the magnetic observatories may possibly be determined with greater precision than by any yet known method.

So long ago as 1839 Professor Morse suggested some experiments for the determination of Longitudes; and in June 1844 the difference of longitude between Washington and Baltimore was determined by electric means under his direction. Two persons were stationed at these two towns, with clocks carefully adjusted to the respective spots; and a telegraphic signal gave the means of comparing the two clocks at a given instant. In 1847 the relative longitudes of New York, Philadelphia, and

Washington were determined by means of the electric telegraph by Messrs. Keith, Walker, and Loomis.

NON-INTERFERENCE OF GALVANIC WAVES ON THE SAME WIRE.

One of the most remarkable facts in the economy of the telegraph is, that the line, when connected with a battery in action, propagates the hydro-galvanic waves in either direction without interference. As several successive syllables of sound may set out in succession from the same place, and be on their way at the same time, to a listener at a distance, so also, where the telegraph-line is long enough, several waves may be on their way from the signal station before the first one reaches the receiving station; two persons at a distance may pronounce several syllables at the same time, and each hear those emitted by the other. So, on a telegraph-line of two or three thousand miles in length in the air, and the same in the ground, two operators may at the same instant commence a series of several dots and lines, and each receive the other's writings, though the waves have crossed each other on the way.

EFFECT OF LIGHTNING UPON THE ELECTRIC TELEGRAPH.

In the storm of Sunday April 2, 1848, the lightning had a very considerable effect on the wires of the electric telegraph, particularly on the line of railway eastward from Manchester to Normanton. Not only were the needles greatly deflected, and their power of answering to the handles considerably weakened, but those at the Normanton station were found to have had their poles reversed by some action of the electric fluid in the atmosphere. The damage, however, was soon repaired, and the needles again put in good working order.

ELECTRO-TELEGRAPHIC MESSAGE TO THE STARS.

The electric fluid travels at the mean rate of 20,000 miles in a second under ordinary circumstances; therefore, if it were possible to establish a telegraphic communication with the star 61 Cygni, it would require ninety years to send a message there.

Professor Henderson and Mr. Maclear have fully confirmed the annual parallax of α Centauri to amount to a second of arc, which gives about twenty billions of miles as its distance from our system; a ray of light would arrive from α Centauri to us in little more than three years, and a telegraphic despatch would arrive there in thirty years.

THE ATLANTIC TELEGRAPH.

The telegraphic communication between England and the United States is so grand a conception, that it would be impossible to detail its scientific and mechanical relations within the limits of the present work. All that we shall attempt, therefore, will be to glance at a few of the leading operations.

In the experiments made before the Atlantic Telegraph was finally decided on, 2000 miles of subterranean and submarine telegraphic wires, ramifying through England and Ireland and under the waters of the Irish Sea, were specially connected for the purpose; and through this distance of 2000 miles 250 distinct signals were recorded and printed in one minute.

First, as to the *Cable*. In the ordinary wires by the side of a railway the electric current travels on with the speed of lightning—uninterrupted by the speed of lightning; but when a wire is encased in gutta-percha, or any similar covering, for submersion in the sea, new forces come into play. The electric excitement of the wire acts by induction, through the envelope, upon the particles of water in contact with that envelope, and calls up an electric force of an opposite kind. There are two forces, in fact, pulling against each other through the gutta-percha as a neutral medium,—that is, the electricity in the wire, and the opposite electricity in the film of water immediately surrounding the cable; and to that extent the power of the current in the enclosed wire is weakened. A submarine cable, when in the water, is virtually *a lengthened-out Leyden jar*; it transmits signals while being charged and discharged, instead of merely allowing a stream to flow evenly along it: it is a *bottle* for holding electricity rather than a *pipe* for carrying it; and this has to be filled for every time of using. The wire being carried underground, or through the water, the speed becomes quite measurable, say a thousand miles in a second, instead of two hundred thousand, owing to the retardation by induced or retrograde currents. The energy of the currents and the quality of the wire also affect the speed. Until lately it was supposed that the wire acts only as a *conductor* of electricity, and that a long wire must produce a weaker effect than a short one, on account of the consequent attenuation of the electrical influence; but it is now known that, the cable being a *reservoir* as well as a conductor, its electrical supply is increased in proportion to its length.

The electro-magnetic current is employed, since it possesses a treble velocity of transmission, and realises consequently *a threefold working speed* as compared with simple voltaic electricity. Mr. Wildman Whitehouse has determined by his ingenious apparatus that the speed of the voltaic current might be raised under special circumstances to 1800 miles per second; but that of the induced current, or the electro-magnetic, might be augmented to 6000 miles per second.

Next as to a *Quantity Battery* employed in these investigations. To effect a charge, and transmit a current through some thousand miles of the Atlantic Cable, Mr. Whitehouse had a piece of apparatus prepared consisting of twenty-five pairs of zinc and silver plates about the 20th part of a square inch large, and the pairs so arranged that they would hold a drop of acidulated water or brine between them. On charging this Lilliputian battery by dipping the plates in salt and water, messages were sent from it through a thousand miles of cable with the utmost ease; and not only so,—pair after pair was dropped out from the series, the messages being still sent on with equal facility, until at last only a single pair, charged by one single drop of liquid, was used. Strange to say, with this single pair and single drop distinct signals were effected through the thousand miles of the cable! Each signal was registered at the end of the cable in less than three seconds of time.

The entire length of wire, iron and copper, spun into the cable amounts to 332,500 miles, a length sufficient to engirdle the earth thirteen times. The cable weighs from 19 cwt. to a ton per mile, and will bear a strain of 5 tons.

The *Perpetual Maintenance Battery*, for working the cable at the bottom of the sea, consists of large plates of platinated silver and amalgamated zinc, mounted in cells of gutta-percha. The zinc plates in each cell rest upon a longitudinal bar at the bottom, and the silver plates hang upon a similar bar at the top of the cell; so that there is virtually but a single stretch of silver and a single stretch of zinc in operation. Each of the ten cells contains 2000 square inches of acting surface; and the combination is so powerful, that when the broad strips of copper-plate which form the polar extensions are brought into contact or separated, brilliant flashes are produced, accompanied by a loud crackling sound. The points of large pliers are made red-hot in five seconds when placed between them, and even screws burn with vivid scintillation. The cost of maintaining this magnificent ten-celled Titan battery at work does not exceed a shilling per hour. The voltaic current generated in this battery is not, however, the electric stream to be sent across the Atlantic, but is only the primary power used to call up and stimulate the energy of a more speedy traveller by a complicated apparatus of "Double Induction Coils." Nor is the transmission-current generated in the inner wire of the double induction coil,—and which becomes weakened when it has passed through 1800 or 1900 miles,—set to work to print or record the signals transmitted. This weakened current merely opens and closes the outlet of a fresh battery, which is to do the printing labour. This relay-instrument (as it is called), which consists of a temporary and permanent magnet, is so sensitive an apparatus, that it may be put in action by a fragment of zinc and a sixpence pressed against the tongue.

The attempts to lay the cable in August 1857 failed through stretching it so tightly that it snapped and went to the bottom, at a depth of 12,000 feet, forty times the height of St. Paul's.

This great work was resumed in August 1858; and on the 5th the first signals were received through *two thousand and fifty miles* of the Atlantic Cable. And it is worthy of remark, that just 111 years previously, on the 5th of August 1747, Dr. Watson astonished the scientific world by practically proving that the electric current could be transmitted through a *wire hardly two miles and a half long.*[55]

Miscellanea.

HOW MARINE CHRONOMETERS ARE RATED AT THE ROYAL OBSERVATORY, GREENWICH.

The determination of the Longitude at Sea requires simply accurate instruments for the measurement of the positions of the heavenly bodies, and one or other of the two following,—either perfectly correct watches—or chronometers, as they are now called—or perfectly accurate tables of the lunar motions.

So early as 1696 a report was spread among the members of the Royal Society that Sir Isaac Newton was occupied with the problem of finding the longitude at sea; but the rumour having no foundation, he requested Halley to acquaint the members "that he was not about it."[56] (*Sir David Brewster's Life of Newton.*)

In 1714 the legislature of Queen Anne passed an Act offering a reward of 20,000*l.* for the discovery of the longitude, the problem being then very inaccurately solved for want of good watches or lunar tables. About the year 1749, the attention of the Royal Society was directed to the improvements effected in the construction of watches by John Harrison, who received for his inventions the Copley Medal. Thus encouraged, Harrison continued his labours with unwearied diligence, and produced in 1758 a timekeeper which was sent for trial on a voyage to Jamaica. After 161 days the error of the instrument was only 1^m 5^s, and the maker received from the nation 5000*l.* The Commissioners of the Board of Longitude subsequently required Harrison to construct under their inspection chronometers of a similar nature, which were subjected to trial in a voyage to Barbadoes, and performed with such accuracy, that, after having fully explained the principle of their construction to the commissioners, they awarded him 10,000*l.* more; at the same time Euler of Berlin and the heirs of Mayer of Göttingen received each 3000*l.* for their lunar tables.

The account of the trial of Harrison's watch is very interesting. In April 1766, by desire of the Commissioners of the Board, the Lords of the Admiralty delivered the watch into the custody of the Astronomer-Royal, the Rev. Dr. Nevil Maskelyne. It was then placed at the Royal Observatory at Greenwich, in a box having two different locks, fixed to the floor or wainscot, with a plate of glass in the lid of the box, so that it might be compared as often as convenient with the regulator and the variation set down. The form observed by Mr. Harrison in winding up the watch was exactly followed; and an officer of Greenwich Hospital attended every day, at a stated hour, to see the watch wound up, and its comparison with the regulator entered. A key to one of the locks was kept at the Hospital for the use of the officer, and the other remained at the Observatory for the use of the Astronomer-Royal or his assistant.

The watch was then tried in various positions till the beginning of July; and from thence to the end of February following in a horizontal position with its face upwards.

The variation of the watch was then noted down, and a register was kept of the barometer and thermometer; and the time of comparing the same with the regulator was regularly kept, and attested by the Astronomer-Royal or his assistant and such of the officers as witnessed the winding-up and comparison of the watch.

Under these conditions Harrison's watch was received by the Astronomer-Royal at the Admiralty on May 5, 1766, in the presence of Philip Stephens, Esq., Secretary of the Admiralty; Captain Baillie, of the Royal Hospital, Greenwich; and Mr. Kendal the watchmaker, who accompanied the Astronomer-Royal to Greenwich, and saw the watch started and locked up in the box provided for it. The watch was then compared with the transit clock daily, and wound up in the presence of the officer of Greenwich Hospital. From May 5 to May 17 the watch was kept in a horizontal position with its face upwards; from May 18 to July 6 it was tried—first inclined at an angle of 20° to the horizon, with the face upwards, and the hours 12, 6, 3, and 9, highest successively; then in a vertical position, with the same hours highest in order; lastly, in a horizontal position with the face downwards. From July 16, 1766, to March 4, 1767, it was always kept in a horizontal position with its face upwards, lying upon the same cushion, and in the same box in which Mr. Harrison had kept it in the voyage to Barbadoes.

From the observed transits of the sun over the meridian, according to the time of the regulator of the Observatory, together with the attested comparisons of Mr. Harrison's watch with the transit clock, the watch was found too fast on several days as follows:

			h.	m.	s.
1766.	May 6	too fast	0	0	16·2
	May 17	"	0	3	51·8
	July 6	"	0	14	14·0
	Aug. 6	"	0	23	58·4

	Sept. 17	,,	0	32	15·6
	Oct. 29	,,	0	42	20·9
	Dec. 10	,,	0	54	46·8
1767.	Jan. 21	,,	1	0	28·6
	March 4	,,	1	11	23·0

From May 6, which was the day after the watch arrived at the Royal Observatory, to March 4, 1767, there were six periods of six weeks each in which the watch was tried in a horizontal position; when the gaining in these several periods was as follows:

During the first 6 weeks	it gained	13^m	20^s,	answering to	3°	20′	of longitude.
In the 2d period of 6 weeks (from Aug. 6 to Sept. 17)	,,	8	17	,,	2	4	,,
In the 3d period (from Sept. 17 to Oct. 29)	,,	10	5	,,	2	31	,,
In the 4th period (from Oct. 29 to Dec. 20)	,,	12	26	,,	3	6	,,
In the 5th period (from Dec. 20 to Jan. 21)	,,	5	42	,,	1	25	,,
In the 6th period (from Jan. 21 to Mar. 4)	,,	10	54	,,	2	43	,,

It was thence concluded that Mr. Harrison's watch could not be depended upon to keep the longitude within a West-India voyage of six weeks, nor to keep the longitude within half a degree for more than a fortnight; and that it must be kept in a place where the temperature was always some degrees above freezing.57 (However, Harrison's watch, which was made by Mr. Kendal subsequently, succeeded so completely, that after it had been round the world with Captain Cook, in the years 1772–1775, the second 10,000*l.* was given to Harrison.)

In the Act of 12th Queen Anne, the comparison of chronometers was not mentioned in reference to the Observatory duties; but after this time they became a serious charge upon the Observatory, which, it must be admitted, is by far the best place to try chronometers: the excellence of the instruments, and the frequent observations of the heavenly bodies over the meridian, will always render the rate of going of the Observatory clock better known than can be expected of the clock in most other places.

After Mr. Harrison's watch was tried, some watches by Earnshaw, Mudge, and others, were rated and examined by the Astronomer-Royal.

At the Royal Observatory, Greenwich, there are frequently above 100 chronometers being rated, and there have been as many as 170 at one time. They are rated daily by two observers, the process being as follows. At a certain time every day two assistants in charge repair to the chronometer-room, where is a time-piece set to true time; one winds up each with its own key, and the second follows after some little time and verifies the fact that each is wound. One assistant then looks at each watch in succession, counting the beats of the clock whilst he compares the chronometer by the eye; and in the course of a few seconds he calls out the second shown by the chronometer when the clock is at a whole minute. This number is entered in a book by the other assistant, and so on till all the chronometers are compared. Then the assistants change places, the second comparing and the first writing down. From these daily comparisons the daily rates are deduced, by which the goodness of the watch is determined. The errors are of two classes—that of general bad workmanship, and that of over or under correction for temperature. In the room is an apparatus in which the watch may be continually kept at temperatures exceeding 100° by artificial heat; and outside the window of the room is an iron cage, in which they are subjected to low temperatures. The very great care taken with all chronometers sent to the Royal Observatory, as well as the perfect impartiality of the examination which each receives, afford encouragement to their manufacture, and are of the utmost importance to the safety and perfection of navigation.

We have before us now the Report of the Astronomer-Royal on the Rates of Chronometers in the year 1854, in which the following are the successive weekly sums of the daily rates of the first there mentioned:

Week ending		secs.
Jan. 21, loss in the week		2•2
" 28	"	4•0
Feb. 4	"	1•1
" 11	"	5•0
" 18	"	4•9
" 25	"	5•5
Mar. 4	"	6•0
" 11	"	6•0
" 18	"	1•5
" 25	"	4•5
Apr. 1	"	4•0
" 8	"	1•5

”	15, gain in the week	0•4
Apr. 22,	”	2•6
”	29, loss in the week	1•4
May 6	”	2•1
” 13	”	3•0
” 20	”	5•1
” 27	”	3•3
June 3	”	2•8
” 10	”	1•8
” 17	”	2•0
” 24	”	3•0
July 1	”	2•5
” 8	”	1•2

Till February 4 the watch was exposed to the external air outside a north window; from February 5 to March 4 it was placed in the chamber of a stove heated by gas to a moderate temperature; and from April 29 to May 20 it was placed in the chamber when heated to a high temperature.

The advance in making chronometers since Harrison's celebrated watch was tried at the Royal Observatory, more than ninety years since, may be judged by comparing its rates with those above.

GEOMETRY OF SHELLS.

There is a mechanical uniformity observable in the description of shells of the same species which at once suggests the probability that the generating figure of each increases, and that the spiral chamber of each expands itself, according to some simple geometrical law common to all. To the determination of this law the operculum lends itself, in certain classes of shells, with remarkable facility. Continually enlarged by the animal, as the construction of its shell advances so as to fill up its mouth, the operculum measures the progressive widening of the spiral chamber by the progressive stages of its growth.

* * * * *

The animal, as he advances in the construction of his shell, increases continually his operculum, so as to adjust it to his mouth. He increases it, however, not by additions made at the same time all round its margin, but by additions made only on one side of it at once. One edge of the operculum thus remains unaltered as it is advanced into each new position, and placed in a newly-formed section of the chamber similar to the last but greater than it.

That the same edge which fitted a portion of the first less section should be capable of adjustment so as to fit a portion of the next similar but greater section, supposes a geometrical provision in the curved form of the chamber of great complication and difficulty. But God hath bestowed upon this humble architect the practical skill of the learned geometrician; and he makes this provision with admirable precision in that curvature of the logarithmic spiral which he gives to the section of the shell. This curvature obtaining, he has only to turn his operculum slightly round in its own place, as he advances it into each newly-formed portion of his chamber, to adapt one margin of it to a new and larger surface and a different curvature, leaving the space to be filled up by increasing the operculum wholly on the outer margin.

* * * * *

Why the Mollusks, who inhabit turbinated and discoid shells, should, in the progressive increase of their spiral dwellings, affect the peculiar law of the logarithmic spiral, is easily to be understood. Providence has subjected the instinct which shapes out each to a rigid uniformity of operation.—*Professor Mosely: Philos. Trans.* 1838.

HYDRAULIC THEORY OF SHELLS.

How beautifully is the wisdom of God developed in shaping out and moulding shells! and especially in the particular value of the constant angle which the spiral of each species of shell affects,—a value connected by a necessary relation with the economy of the material of each, and with its stability and the conditions of its buoyancy. Thus the shell of the *Nautilus Pompilius* has, hydrostatically, an A-statical surface. If placed with any portion of its surface upon the water, it will immediately turn over towards its smaller end, and rest only on its mouth. Those conversant with the theory of floating bodies will recognise in this an interesting property.—*Ibid.*

SERVICES OF SEA-SHELLS AND ANIMALCULES.

Dr. Maury is disposed to regard these beings as having much to do in maintaining the harmonies of creation, and the principles of the most admirable compensation in the system of oceanic circulation. "We may even regard them as regulators, to some extent, of climates in parts of the earth far removed from their presence. There is something suggestive both of the grand and the beautiful in the idea that while the insects of the sea are building up their coral islands in the perpetual summer of the tropics, they are also engaged in dispensing warmth to distant parts of the earth, and in mitigating the severe cold of the polar winter."

DEPTH OF THE PRIMEVAL SEAS.

Professor Forbes, in a communication to the Royal Society, states that not only the colour of the shells of existing mollusks ceases to be strongly marked at considerable depths, but also that well-defined patterns are, with very few and slight exceptions, presented only by testacea inhabiting the littoral, circumlittoral, and median zones. In the Mediterranean, only one in eighteen of the shells taken from below 100 fathoms exhibit any markings of colour, and even the few that do so are questionable inhabitants of those depths. Between 30 and 35 fathoms, the proportion of marked to plain shells is rather less than one in three; and between the margin and two fathoms the striped or mottled species exceed one-half of the total number. In our own seas, Professor Forbes observes that testacea taken from below 100 fathoms, even when they are individuals of species vividly striped or banded in shallower zones, are quite white or colourless. At between 60 and 80 fathoms, striping and banding are rarely presented by our shells, especially in the northern provinces; from 50 fathoms, shallow bands, colours, and patterns, are well marked. *The relation of these arrangements of colour to the degree of light penetrating the different zones of depth* is a subject well worthy of minute inquiry.

NATURAL WATER-PURIFIERS.

Mr. Warrington kept for a whole year twelve gallons of water in a state of admirably balanced purity by the following beautiful action:

In the tank, or aquarium, were two gold fish, six water-snails, and two or three specimens of that elegant aquatic plant *Valisperia sporalis*, which, before the introduction of the water-snails, by its decayed leaves caused a growth of slimy mucus, and made the water turbid and likely to destroy both plants and fish. But under the improved arrangement the slime, as fast as it was engendered, was consumed by the water-snails, which reproduced it in the shape of young snails, which furnished a succulent food to the fish. Meanwhile the *Valisperia* plants absorbed the carbonic acid exhaled by the respiration of their companions, fixing the carbon in their growing stems and luxuriant blossoms, and refreshing the oxygen (during sunshine in visible little streams) for the respiration of the snails and the fish. The spectacle of perfect equilibrium thus simply maintained between animal, vegetable, and inorganic activity, was strikingly beautiful; and such means might possibly hereafter be made available on a large scale for keeping tanked water sweet and clean.—*Quarterly Review*, 1850.

HOW TO IMITATE SEA-WATER.

The demand for Sea-water to supply the Marine Aquarium—now to be seen in so many houses—induced Mr. Gosse to attempt the manufacture of Sea-water, more especially as the constituents are well known. He accordingly took Scheveitzer's analysis of Sea-water for his guide. In one thousand grains of sea-water taken off Brighton, it gave: water, 964•744; chloride of sodium, 27•059; chloride of magnesium, 3•666; chloride of potassium, 9•755; bromide of magnesium, 0•29; sulphate of magnesia, 2•295; sulphate of lime, 1•407; carbonate of lime, 0•033: total, 999•998. Omitting the bromide of magnesium, the carbonate of lime, and the sulphate of lime, as being very small quantities, the component parts were reduced to common salt, 3½ oz.; Epsom salts, ¼ oz.; chloride of magnesium, 200 grains troy; chloride of potassium, 40 grains troy; and four quarts of water. Next day the mixture was filtered through a sponge into a glass jar, the bottom covered with shore-pebbles and fragments of stone and fronds of green sea-weed. A coating of green spores was soon deposited on the sides of the glass, and bubbles of oxygen were copiously thrown off every day under the excitement of the sun's light. In a week Mr. Gosse put in species of *Actinia Bowerbankia*, *Cellularia*, *Serpula*, &c. with some red sea-weeds; and the whole throve well.

VELOCITY OF IMPRESSIONS TRANSMITTED TO THE BRAIN.

Professor Helmholtz of Königsberg has, by the electro-magnetic method,[58] ascertained that the intelligence of an impression made upon the ends of the nerves in communication with the skin is transmitted to the brain with a velocity of about 195 feet per second. Arrived at the brain, about one-tenth of a second passes before the will is able to give the command to the nerves that certain muscles shall execute a certain motion, varying in persons and times. Finally, about 1/100th of a second passes after the receipt of the command before the muscle is in activity. In all, therefore, from the excitation of the sensitive nerves till the moving of the muscle, 1¼ to 2/10ths of a second are consumed. Intelligence from the great toe arrives about 1/30th of a second later than from the ear or the face.

Thus we see that the differences of time in the nervous impressions, which we are accustomed to regard as simultaneous, lie near our perception. We are taught by astronomy that, on account of the time taken to propagate light, we now see what has occurred in the fixed stars years ago; and that, owing to the time required for the transmission of sound, we hear after we see is a matter of daily experience. Happily the distances to be traversed by our sensuous perceptions before they reach the brain are so short that we do not observe their influence, and are therefore unprejudiced in our practical interest. With an ordinary whale the case is perhaps more dubious; for in all probability the animal does not feel a wound near its tail until a second after it has been inflicted, and requires another second to send the command to the tail to defend itself.

PHOTOGRAPHS ON THE RETINA.

The late Rev. Dr. Scoresby explained with much minuteness and skill the varying phenomena which presented themselves to him after gazing intently for some time on strongly-illuminated objects,—as the sun, the moon, a red or orange or yellow wafer on a strongly-contrasted ground, or a dark object seen in a bright field. The doctor explained, upon removing the eyes from the object, the early appearance of the picture or image which had been thus "photographed on the Retina," with the photochromatic changes which the picture underwent while it still retained its general form and most strongly-marked features; also, how these pictures, when they had almost faded away, could at pleasure, and for a considerable time, be renewed by rapidly opening and shutting the eyes.

DIRECT EXPLORATION OF THE INTERIOR OF THE EYE.

Dr. S. Wood of Cincinnati states, that by means of a small double convex lens of short focus held near the eye,—that organ looking through it at a candle twelve or fifteen feet distant,—there will be perceived a large luminous disc, covered with dark and light spots and dark streaks, which, after a momentary confusion, will settle down into an unchanging picture, which picture is composed of the organs or internal parts of the eye. The eye is thus enabled to view its own internal organisation, to have a beautiful exhibition of the vessels of the cornea, of the distribution of the lachrymas secretions in the act of winking, and to see into the nature and cause of *muscæ volitantes*.

NATURE OF THE CANDLE-FLAME.

M. Volger has subjected this Flame to a new analysis.

He finds that the so-called *flame-bud*, a globular blue flaminule, is first produced at the summit of the wick: this is the result of the combustion of carbonic oxide, hydrogen, and carbon, and is surrounded by a reddish-violet halo, the *veil*. The increased heat now gives rise to the actual flame, which shoots forth from the expanding bud, and is then surrounded at its inferior portion only by the latter. The interior consists of a dark gaseous cone, containing the immediate products of the decomposition of the fatty acids, and surrounded by another dark hollow cone, the *inner cap*. Here we already meet with carbon and hydrogen, which have resulted from the process of decomposition; and we distinguish this cone from the inner one by its yielding soot. The *external cap* constitutes the most luminous portion of the flame, in which the hydrogen is consumed and the carbon rendered incandescent. The surrounding portion is but slightly luminous, deposits no soot, and in it the carbon and hydrogen are consumed.—*Liebig's Annual Report.*

HOW SOON A CORPSE DECAYS.

Mr. Lewis, of the General Board of Health, from his examination of the contents of nearly 100 coffins in the vaults and catacombs of London churches, concludes that the complete decomposition of a corpse, and its resolution into its ultimate elements, takes place in a leaden coffin with extreme slowness. In a wooden coffin the remains, with the exception of the bones, vanish in from two to five years. This period depends upon the quality of the wood, and the free access of air to the coffins. But in leaden coffins, 50, 60, 80, and even 100 years are required to accomplish this. "I have opened," says Mr. Lewis, "a coffin in which the corpse had been placed for nearly a century; and the ammoniacal gas formed dense white fumes when brought in contact with hydrochloric-acid gas, and was so powerful that the head could not remain in it for more than a few seconds at a time." To render the human body perfectly inert after death, it should be placed in a light wooden coffin, in a pervious soil, from five to eight feet deep.

MUSKET-BALLS FOUND IN IVORY.

The Ceylon sportsman, in shooting elephants, aims at a spot just above the proboscis. If he fires a little too low, the ball passes into the tusk-socket, causing great pain to the animal, but not endangering its life; and it is immediately surrounded by osteo-dentine. It has often been a matter of wonder how such bodies should become completely imbedded in the substance of the tusk, sometimes without any visible aperture; or how leaden bullets become lodged in the solid centre of a very large tusk without having been flattened, as they are found by the ivory-turner.

The explanation is as follows: A musket-ball aimed at the head of an elephant may penetrate the thin bony socket and the thinner ivory parietes of the wide conical pulp-cavity occupying the inserted base of the tusk; if the projectile force be there spent, the ball will gravitate to the opposite and lower side of the pulp-cavity. The pulp becomes inflamed, irregular calcification ensues, and osteo-dentine is formed around the ball. The pulp then resumes its healthy state and functions, and coats the osteo-dentine enclosing the ball, together with the root of the conical cavity into which the mass projects, with layers of normal ivory. The hole formed by the ball is soon replaced, and filled up by osteo-dentine, and coated with cement. Meanwhile, by the continued progress of growth, the enclosed ball is pushed forward to the middle of the solid tusk; or if the elephant be young, the ball may be carried forward by growth and wear of the tusk until its base has become the apex, and become finally exposed and discharged by the continual abrasion to which the apex of the tusk is subjected.—*Professor Owen.*

NATURE OF THE SUN.

To the article at pp. 59–60 should be added the result obtained by Dr. Woods of Parsonstown, and communicated to the *Philosophical Magazine* for July 1854. Dr. Woods, from photographic experiment, has no doubt that the light from the centre of flame acts more energetically than that from the edge on a surface capable of receiving its impression; and that light from a luminous solid body acts equally powerfully from its centre or its edges: wherefore Dr. Woods concludes that, as the sun affects a sensitive plate similarly with flame, it is probable its light-producing portion is of a similar nature.

Note to " IS THE HEAT OF THE SUN DECREASING? " *at page 65.*—Dr. Vaughan of Cincinnati has stated to the British Association: "From a comparison of the relative intensity of solar, lunar, and artificial light, as determined by Euler and Wollaston, it appears that the rays of the sun have an illuminating power equal to that of 14,000 candles at a distance of one foot, or of 3500,000000,000000,000000,000000 candles at a distance of 95,000,000 miles. It follows that the amount of light which flows from the solar orb could be scarcely produced by the daily combustion of 200 globes of tallow, each equal to the earth in magnitude. A sphere of combustible matter much larger than the sun itself should be consumed every ten years in maintaining its wonderful brilliancy; and its atmosphere, if pure oxygen, would be expended before a few days in supporting so great a conflagration. An illumination on so vast a scale could be kept up only by the inexhaustible magazine of ether disseminated through space, and ever ready to manifest its luciferous properties on large spheres, whose attraction renders it sufficiently dense for the play of chemical affinity. Accordingly suns derive the power of shedding perpetual light, not from their chemical constitution, but from their immense mass and their superior attractive power."

PLANETOIDS.

Name.	Date of Discovery.	Discoverer.	Place of Discovery.	No. discovered by each astronomer.

Mercury, Mars, Venus, Jupiter, Earth, Saturn	Known to the ancients.	...	...	—
Uranus	1781, March 13	W. Herschel	Bath	—
Neptune[59]	1846, Sept. 23	Galle	Berlin	—
1 Ceres	1801, Jan. 1	Piazzi	Palermo	1
2 Pallas	1802, March 28	Olbers	Bremen	1
3 Juno	1804, Sept. 1	Harding	Lilienthal	1
4 Vesta	1807, March 29	Olbers	Bremen	2
5 Astræa	1845, Dec. 8	Encke	Driesen	1
6 Hebe	1847, July 1	Encke	Driesen	2
7 Iris	1847, August 13	Hind	London	1
8 Flora	1847, Oct. 18	Hind	London	2
9 Metis	1848, April 25	Graham	Markree	1
10 Hygeia	1849, April 12	Gasperis	Naples	1

11 Parthenope	1850, May 11	Gasperis	Naples	2
12 Victoria	1850, Sept. 13	Hind	London	3
13 Egeria	1850, Nov. 2	Gasperis	Naples	3
14 Irene	1851, May 19	Hind	London	4
15 Eunomia	1851, July 29	Gasperis	Naples	4
16 Psyche	1852, March 17	Gasperis	Naples	5
17 Thetis	1852, April 17	Luther	Bilk	1
18 Melpomene	1852, June 24	Hind	London	5
19 Fortuna	1852, August 22	Hind	London	6
20 Massilia	1852, Sept. 19	Gasperis	Naples	6
21 Lutetia	1852, Nov. 15	Goldschmidt	Paris	1
22 Calliope	1852, Nov. 16	Hind	London	7
23 Thalia	1852, Dec. 15	Hind	London	8
24 Themis	1853, April 5	Gasperis	Naples	7

25 Phocea	1853, April 6	Chacornac	Marseilles	1
26 Proserpine	1853, May 5	Luther	Bilk	2
27 Euterpe	1853, Nov. 8	Hind	London	9
28 Bellona	1854, March 1	Luther	Bilk	3
29 Amphitrite	1854, March 1	Marth	London	1
30 Urania	1854, July 22	Hind	London	10
31 Euphrosyne	1854, Sept. 1	Furguson	Washington	1
32 Pomona	1854, Oct. 26	Goldschmidt	Paris	2
33 Polyhymnia	1854, Oct. 28	Chacornac	Paris	2
34 Circe	1855, April 6	Chacornac	Paris	3
35 Leucothea	1855, April 19	Luther	Bilk	4
36 Atalante	1855, Oct. 5	Goldschmidt	Paris	3
37 Fides	1855, Oct. 5	Luther	Bilk	5
38 Leda	1856, Jan. 12	Chacornac	Paris	4

39 Lætitia	1856, Feb. 8	Chacornac	Paris	5
40 Harmonia	1856, March 31	Goldschmidt	Paris	4
41 Daphne	1856, May 22	Goldschmidt	Paris	5
42 Isis	1856, May 23	Pogson	Oxford	1
43 Ariadne	1857, April 15	Pogson	Oxford	2
44 Nysa	1857, May 27	Goldschmidt	Paris	6
45 Eugenia	1857, June 28	Goldschmidt	Paris	7
46 Hastia	1857, August 16	Pogson	Oxford	3
47 Aglaia	1857, Sept. 15	Luther	Bilk	6
48 Doris	1857, Sept. 19	Goldschmidt	Paris	8
49 Pales	1857, Sept. 19	Goldschmidt	Paris	9
50 Virginia	1857, Oct. 4	Furguson	Washington	2
51 Nemausa	1858, Jan. 22	Laurent	Nismes	1
52 Europa	1858, Feb. 6	Goldschmidt	Paris	10

53 Calypso	1858, April 8	Luther	Bilk	7
54 Alexandra	1858, Sept. 11	Goldschmidt	Paris	11
55 (Not named)	1858, Sept. 11	Searle	Albany	1

THE COMET OF DONATI.

While this sheet was passing through the press, the attention of astronomers, and of the public generally, was drawn to the fact of the above Comet passing (on Oct. 18) within nine millions of miles of the planet Venus, or less than 9/100ths of the earth's distance from the Sun. "And (says Mr. Hind, the astronomer), it is obvious that if the comet had reached its least distance from the sun a few days earlier than it has done, the planet might have passed through it; and I am very far from thinking that close proximity to a comet of this description would be unattended with danger. The inhabitants of Venus will witness a cometary spectacle far superior to that which has recently attracted so much attention here, inasmuch as the tail will doubtless appear twice as long from that planet as from the earth, and the nucleus proportionally more brilliant."

This Comet was first discovered by Dr. G. B. Donati, astronomer at the Museum of Florence, on the evening of the 2d of June, in right ascension 141° 18', and north declination 23° 47', corresponding to a position near the star Leonis. Previous to this date we had no knowledge of its existence, and therefore it was not a predicted comet; neither is it the one last observed in 1556. At the date of discovery it was distant from the earth 228,000,000 of miles, and was an excessively faint object in the largest telescopes.

The tail, from October 2 to 16, when the comet was most conspicuous, appears to have maintained an average length of at least 40,000,000 miles, subtending an angle varying from 30° to 40°. The dark line or space down the centre, frequently remarked in other great comets, was a striking characteristic in that of Donati. The nucleus, though small, was intensely brilliant in powerful instruments, and for some time bore high magnifiers to much greater advantage than is usual with these objects. In several respects this comet resembled the famous ones of 1744, 1680, and 1811, particularly as regards the signs of violent agitation going on in the vicinity of the nucleus, such as the appearance of luminous jets, spiral offshoots, &c., which rapidly emanated from the planetary point and as quickly lost themselves in the general nebulosity of the head.

On the 5th Oct. the most casual observer had an opportunity of satisfying himself as to the accuracy of the mathematical theory of the motions of comets in the near approach of the nucleus of Donati's to Arcturus, the principal star in the constellation Bootes. The circumstance of the appulse was very nearly as predicted by Mr. Hind.

The comet, according to the investigations by M. Loewy, of the Observatory of Vienna, arrived at its least distance from the sun a few minutes after eleven o'clock on the morning of the 30th of September; its longitude, as seen from the sun at this time, being 36° 13', and its distance from him 55,000,000 miles. The longer diameter of its orbit is 184 times that of the earth's, or 35,100,000,000 miles; yet this is considerably less than 1/1000th of the distance of the nearest fixed star. As an illustration, let any one take a half-sheet of note-paper, and marking a circle with a sixpence in one corner of it, describe therein our solar system, drawing the orbits of the earth and the inferior planets as small as he can by the aid of a magnifying-glass. If the circumference of the sixpence stands for the orbit of Neptune, then an oval filling the page will fairly represent the orbit of Donati's comet; and if the paper be laid upon the pavement under the west door of St. Paul's Cathedral, London, the length of that edifice will inadequately represent the distance of the nearest fixed star. The time of revolution resulting from Mr. Loewy's calculations is 2495 years, which is about 500 years less than that of the comet of 1811 during the period it was visible from the earth.

That the comet should take more than 2000 years to travel round the above page of note-paper is explained by its great diminution of speed as it recedes from the sun. At its perihelion it travelled at the rate of 127,000 miles an hour, or more than twice as fast as the earth, whose motion is about 1000 miles a minute. At its aphelion, however, or its greatest distance from the sun, the comet is a very slow body, sailing at the rate of 480 miles an hour, or only eight times the speed of a railway express. At this pace, were it to travel onward in a straight line, the lapse of a million of years would find it still travelling half way between our sun and the nearest fixed star.

As this comet last visited us between 2000 and 2495 years since, we know that its appearance was at an interesting period of the world's history. It might have terrified the Athenians into accepting the bloody code of Draco. It might have announced the destruction of Nineveh, or of Babylon, or the capture of Jerusalem by Nebuchadnezzar. It might have been seen by the expedition which sailed round Africa in the reign of Pharaoh Necho. It might have given interest to the foundation of the

Pythian games. Within the probable range of its last visitation are comprehended the whole of the great events of the history of Greece; and among the spectators of the comet may have been the so-called sages of Greece and even the prophets of Holy Writ: Thales might have attempted to calculate its return, and Jeremiah might have tried to read its warning.—*Abridged from a Communication from Mr. Hind to the Times, and from a Leader in that Journal.*

FOOTNOTES

1 From a photograph, with figures, to show the relative size of the tube aperture.

2 Weld's *History of the Royal Society*, vol. ii. p. 188.

3 Dr. Whewell (*Bridgewater Treatise*, p. 266) well observes, that Boyle and Pascal are to hydrostatics what Galileo is to mechanics, and Copernicus, Kepler, and Newton are to astronomy.

4 The Rev. Mr. Turnor recollects that Mr. Jones, the tutor, mentioned, in one of his lectures on optics, that the reflecting telescope belonging to Newton was then lodged in the observatory over the gateway; and Mr. Turnor thinks that he once saw it, with a finder affixed to it.

5 The story of the dog "Diamond" having caused the burning of certain papers is laid in London, and in Newton's later years. In the notes to Maude's *Wenleysdale*, a person then living (1780) relates, that Sir Isaac being called out of his study to a contiguous room, a little dog, called Diamond, the constant but incurious attendant of his master's researches, happened to be left among the papers, and by a fatality not to be retrieved, as it was in the latter part of Sir Isaac's days, threw down a lighted candle, which consumed the almost finished labour of some years. Sir Isaac returning too late but to behold the dreadful wreck, rebuked the author of it with an exclamation (*ad sidera palmas*), "O Diamond! Diamond! thou little knowest the mischief done!" without adding a single stripe. M. Biot gives this fiction as a true story, which happened some years after the publication of the *Principia*; and he characterises the accident as having deprived the sciences forever of the fruit of so much of Newton's labours.—Brewster's *Life*, vol. ii. p. 139, note. Dr. Newton remarks, that Sir Isaac never had any communion with dogs or cats; and Sir David Brewster adds, that the view which M. Biot has taken of the idle story of the dog Diamond, charged with fire-raising among Newton's manuscripts, and of the influence of this accident upon the mind of their author, is utterly incomprehensible. The fiction, however, was turned to account in giving colour to M. Biot's misrepresentation.

6 Bohn's edition.

7 When at Pisa, many years since, Captain Basil Hall investigated the origin and divergence of the tower from the perpendicular, and established completely to his own satisfaction that it had been built from top to bottom originally just as it now stands. His reasons for thinking so were, that the line of the tower, on that side towards which it leans, has not the same curvature as the line on the opposite, or what may be called the upper side. If the tower had been built upright, and then been made to incline over, the line of the wall on that side towards which the inclination was given would be more or less concave in that direction, owing to the nodding or "swagging over" of the top, by the simple action of gravity acting on a very tall mass of masonry, which is more or less elastic when placed in a sloping position. But the contrary is the fact; for the line of wall on the side towards which the tower leans is decidedly more convex than the opposite side. Captain Hall had therefore no doubt whatever that the architect, in rearing his successive courses of stones, gained or stole a little at each layer, so as to render his work less and less overhanging as he went up; and thus, without betraying what he was about, really gained stability.—See *Patchwork*.

8 Lord Bacon proposed that, in order to determine whether the gravity of the earth arises from the gravity of its parts, a clock-pendulum should be swung in a mine, as was recently done at Harton colliery by the Astronomer-Royal.

When, in 1812, Ampère noted the phenomena of the pendulum, and showed that its movement was produced only when the eye of the observer was fixed on the instrument, and endeavoured to prove thereby that the motion was due to a play of the muscles, some members of the French Academy objected to the consideration of a subject connected to such an extent with superstition.

9 This curious fact was first recorded by Pepys, in his *Diary*, under the date 31st of July 1665.

10 The result of these experiments for ascertaining the variation of the gravity at great depths, has proved beyond doubt that the attraction of gravitation is increased at the depth of 1250 feet by 1/19000 part.

11 See the account of Mr. Baily's researches (with two illustrations) in *Things not generally Known*, p. vii., and "Weight of the Earth," p. 16.

12 Fizeau gives his result in leagues, reckoning twenty-five to the equatorial degree. He estimates the velocity of light at 70,000 such leagues, or about 210,000 miles in the second.

13 See *Things not generally Known*, p. 88.

14 Some time before the first announcement of the discovery of sun-painting, the following extract from Sir John Herschel's *Treatise on Light*, in the *Encyclopædia Metropolitana*, appeared in a popular work entitled *Parlour Magic*: "Strain a piece of paper or linen upon a wooden frame, and sponge it over with a solution of nitrate of silver in water; place it behind a painting upon glass, or a stained window-pane, and the light, traversing the painting or figures, will produce a copy of it upon the prepared paper or linen; those parts in which the rays were least intercepted being the shadows of the picture."

15 In his book on Colours, Mr. Doyle informs us that divers, if not all, essential oils, as also spirits of wine, when shaken, "have a good store of bubbles, which appear adorned with various and lively colours." He mentions also that bubbles of soap and turpentine exhibit the same colours, which "vary according to the incidence of the sight and the position of the eye;" and he had seen a glass-blower blow bubbles of glass which burst, and displayed "the varying colours of the rainbow, which were exceedingly vivid."

16 The original idea is even attributed to Copernicus. M. Blundevile, in his *Treatise on Cosmography*, 1594, has the following passage, perhaps the most distinct recognition of authority in our language: "How prooue (prove) you that there is but one world? By the authoritie of Aristotle, who saieth that if there were any other world out of this, then the earth of that world would mooue (move) towards the centre of this world," &c.

Sir Isaac Newton, in a conversation with Conduitt, said he took "all the planets to be composed of the same matter with the earth, viz. earth, water, and stone, but variously concocted."

17 Sir William Herschel ascertained that our solar system is advancing towards the constellation Hercules, or more accurately to a point in space whose right ascension is 245° 52' 30", and north polar distance 40° 22'; and that the quantity of this motion is such, that to an astronomer placed in Sirius, our sun would appear to describe an arc of little more than *a second* every year.—*North-British Review*, No. 3.

18 See M. Arago's researches upon this interesting subject, in *Things not generally Known*, p. 4.

19 This eloquent advocacy of the doctrine of "More Worlds than One" (referred to at p. 51) is from the author's valuable *Outlines of Astronomy*.

20 Professor Challis, of the Cambridge Observatory, directing the Northumberland telescope of that institution to the place assigned by Mr. Adams's calculations and its vicinity on the 4th and 12th of August 1846, saw the planet on both those days, and noted its place (among those of other stars) for re-observation. He, however, postponed

the *comparison* of the places observed, and not possessing Dr. Bremiker's chart (which would at once have indicated the presence of an unmapped star), remained in ignorance of the planet's existence as a visible object till the announcement of such by Dr. Galle.

21 For several interesting details of Comets, see "Destruction of the World by a Comet," in *Popular Errors Explained and Illustrated*, new edit. pp. 165–168.

22 The letters of Sir Isaac Newton to Dr. Bentley, containing suggestions for the Boyle Lectures, possess a peculiar interest in the present day. "They show" (says Sir David Brewster) "that the *nebular hypothesis*, the dull and dangerous heresy of the age, is incompatible with the established laws of the material universe, and that an omnipotent arm was required to give the planets their positions and motions in space, and a presiding intelligence to assign to them the different functions they had to perform."—*Life of Newton*, vol. ii.

23 The constitution of the nebulæ in the constellation of Orion has been resolved by this instrument; and by its aid the stars of which it is composed burst upon the sight of man for the first time.

24 Several specimens of Meteoric Iron are to be seen in the Mineralogical Collection in the British Museum.

25 *Life of Sir Isaac Newton*, vol. i. p. 62.

26 *Description of the Monster Telescope*, by Thomas Woods, M.D. 4th edit. 1851.

27 This instrument also discovered a multitude of new objects in the moon; as a mountainous tract near Ptolemy, every ridge of which is dotted with extremely minute craters, and two black parallel stripes in the bottom of Aristarchus. Dr. Robinson, in his address to the British Association in 1843, stated that in this telescope a building the size of the Court-house at Cork would be easily visible on the lunar surface.

28 Mr. Hopkins supports his Glacial Theory by regarding the *Waves of Translation*, investigated by Mr. Scott Russell, as furnishing a sufficient moving power for the transportation of large rounded boulders, and the formation of drifted gravel. When these waves of translation are produced by the sudden elevation of the surface of the sea, the whole mass of water from the surface to the bottom of the ocean moves onward, and becomes a mechanical agent of enormous power. Following up this view, Mr. Hopkins has shown that "elevations of continental masses of only 50 feet each, and from beneath an ocean having a depth of between 300 and 400 feet, would cause the most powerful divergent waves, which could transport large boulders to great distances."

29 It is scarcely too much to say, that from the collection of specimens of building-stones made upon this occasion, and first deposited in a house in Craig's Court, Charing Cross, originated, upon the suggestion of Sir Henry Delabeche, the magnificent Museum of Practical Geology in Jermyn Street; one of the most eminently practical institutions of this scientific age.

30 Mr. R. Mallet, F.R.S., and his son Dr. Mallet, have constructed a seismographic map of the world, with seismic bands in their position and relative intensity; and small black discs to denote volcanoes, femaroles, and soltataras, and shades indicating the areas of subsidence.

31 It has been computed that the shock of this earthquake pervaded an area of 700,000 miles, or the twelfth part of the circumference of the globe. This dreadful shock lasted only five minutes; and nearly the whole of the population being within the churches (on the feast of All Saints), no less than 30,000 persons perished by the fall of these edifices.—See *Daubeny on Volcanoes*; *Translator's note, Humboldt's Cosmos*.

32 Mr. Murray mentions, on the authority of the Rev. Dr. Robinson, of the Observatory at Armagh, that a rough diamond with a red tint, and valued by Mr. Rundell at twenty guineas, was found in Ireland, many years since, in the bed of a brook flowing through the county of Fermanagh.

33 The use of malachite in ornamental work is very extensive in Russia. Thus, to the Great Exhibition of 1851 were sent a pair of folding-doors veneered with malachite, 13 feet high, valued at 600*l*.; malachite cases and pedestals from 1500*l*. to 3000*l*. a-piece, malachite tables 400*l*., and chairs 150*l*. each.

34 Longfellow has written some pleasing lines on "The Fiftieth Birthday of M. Agassiz. May 28, 1857," appended to "The Courtship of Miles Standish," 1858.

35 The *sloth* only deserves its name when it is obliged to attempt to proceed along the ground; when it has any thing which it can lay hold of it is agile enough.

36 Dr. A. Thomson has communicated to *Jameson's Journal*, No. 112, a Description of the Caves in the North Island, with some general observations on this genus of birds. He concludes them to have been indolent, dull, and stupid; to have lived chiefly on vegetable food in mountain fastnesses and secluded caverns.

In the picture-gallery at Drayton Manor, the seat of Sir Robert Peel, hangs a portrait of Professor Owen, and in his hand is depicted the tibia of a Moa.

37 According to the law of correlation, so much insisted on by Cuvier, a superior character implies the existence of its inferiors, and that too in definite proportions and

constant connections; so that we need only the assurance of one character, to be able to reconstruct the whole animal. The triumph of this system is seen in the reconstruction of extinct animals, as in the above case of the Dinornis, accomplished by Professor Owen.

38 Not only at London, but at Paris, Vienna, Berlin, Turin. St. Petersburg, and almost every other capital in Europe; at Liege, Caen, Montpellier, Toulouse, and several other large towns,—wherever, in fact, there are not great local obstacles,—the tendency of the wealthier inhabitants to group themselves to the west is as strongly marked as in the British metropolis. At Pompeii, and other ancient towns, the same thing maybe noticed; and where the local configuration of the town necessitates an increase in a different direction, the moment the obstacle ceases houses spread towards the west.

39 By far the most complete set of experiments on the Radiation of Heat from the Earth's Surface at Night which have been published since Dr. Wells's Memoir *On Dew*, are those of Mr. Glaisher, F.R.S., *Philos. Trans.* for 1847.

40 The author is largely indebted for the illustrations in this new field of research to Lieutenant Maury's valuable work, *The Physical Geography of the Sea*. Sixth edition. Harper, New York; Low, Son, and Co., London.

41 It is the chloride of magnesia which gives that damp sticky feeling to the clothes of sailors that are washed or wetted with salt water.

42 This fraction rests on the assumption that the dilatation of the substances of which the earth is composed is equal to that of glass, that is to say, 1/18000 for 1°. Regarding this hypothesis, see Arago, in the *Annuaire* for 1834, pp. 177–190.

43 Electricity, traversing excessively rarefied air or vapours, gives out light, and doubtless also heat. May not a continual current of electric matter be constantly circulating in the sun's immediate neighbourhood, or traversing the planetary spaces, and exerting in the upper regions of its atmosphere those phenomena of which, on however diminutive a scale, we have yet an unequivocal manifestation in our Aurora Borealis?

44 Could we by mechanical pressure force water into a solid state, an immense quantity of heat would be set free.

45 See Mr. Hunt's popular work, *The Poetry of Science; or, Studies of Physical Phenomena of Nature*. Third edition, revised and enlarged. Bohn, 1854.

46 Canton was the first who in England verified Dr. Franklin's idea of the similarity of lightning and the electric fluid, July 1752.

47 This is mentioned in *Procli Diadochi Paraphrasis Ptolem.*, 1635. (Delambre, *Hist. de l'Astronomie ancienne*.)

48 The first Variation-Compass was constructed, before 1525, by an ingenious apothecary of Seville, Felisse Guillen. So earnest were the endeavours to learn more exactly the direction of the curves of magnetic declination, that in 1585 Juan Jayme sailed with Francisco Gali from Manilla to Acapulco, for the sole purpose of trying in the Pacific a declination instrument which he had invented.—*Humboldt.*

49 Gilbert was surgeon to Queen Elizabeth and James I., and died in 1603. Whewell justly assigns him an important place among the "practical reformers of the physical sciences." He adopted the Copernican doctrine, which Lord Bacon's inferior aptitude for physical research led him to reject.

50 This illustration, it will be seen, does not literally correspond with the details which precede it.

51 Mr. Crosse gave to the meeting a general invitation to Fyne Court; one of the first to accept which was Sir Richard Phillips, who, on his return to Brighton, described in a very attractive manner, at the Sussex Institution, Mr. Crosse's experiments and apparatus; a report of which being communicated to the *Brighton Herald*, was quoted in the *Literary Gazette*, and thence copied generally into the newspapers of the day.

52 These experiments were performed at the expense of the Royal Society, and cost 10*l.* 5*s.* 6*d.* In the Paper detailing the experiments, printed in the 45th volume of the*Philosophical Transactions*, occurs the first mention of Dr. Franklin's name, and of his theory of positive and negative electricity.—*Weld's Hist. Royal Soc.* vol. i. p. 467.

53 In this year Andrew Crosse said: "I prophesy that by means of the electric agency we shall be enabled to communicate our thoughts instantaneously with the uttermost parts of the earth."

54 To which paper the writer is indebted for many of these details.

55 These illustrations have been in the main selected and abridged from papers in the *Companion to the Almanac*, 1858, and the *Penny Cyclopædia*, 2d supp.

56 Newton was, however, much pestered with inquirers; and a Correspondent of the *Gentleman's Magazine*, in 1784, relates that he once had a transient view of a Ms. in Pope's handwriting, in which he read a verified anecdote relating to the above period. Sir Isaac being often interrupted by ignorant pretenders to the discovery of the longitude, ordered his porter to inquire of every stranger who desired admission whether he came about the longitude, and to exclude such as answered in the affirmative. Two lines in Pope's Ms., as the Correspondent recollects, ran thus:

"'Is it about the longitude you come?'The porter asks: 'Sir Isaac's not at home.'"

57 In trying the merits of Harrison's chronometers, Dr. Maskelyne acquired that knowledge of the wants of nautical astronomy which afterwards led to the formation of the Nautical Almanac.

58 A slight electric shock is given to a man at a certain portion of the skin; and he is directed the moment he feels the stroke to make a certain motion, as quickly as he possibly can, with the hands or with the teeth, by which the time-measuring current is interrupted.

59 Through the calculations of M. Le Verrier.